Handbook of the Elements

Handbook of
the Elements

Samuel Ruben

Open Court Publishing Company
La Salle, Illinois 61301

Copyright © 1965, 1967, and 1985 by Samuel Ruben.

First printing 1985.
Second printing 1987.

Printed in the United States of America

Library of Congress Cataloging in Publication Data

Ruben, Samuel
    Handbook of the elements.

    1. Chemical elements.   I. Title
QD466.R78   1985   546   85-18942
ISBN 0-87548-399-2

CU: OpenCourt / OCPublish / Ruben2641 / Copyrt
copyright page for Ruben'sa Handbk of elements
PBT no. 2641
1st run 8/7/87 tvm
Paul Baker Typography, Inc.
fonts 996,571,569

Preface

Handbook of the Elements is a practical reference source that provides essential information on the 108 known chemical elements for students and working scientists alike.

Knowledge about the elements is critical to our understanding of science and the world around us. This edition represents the most up-to-date compilation of information on the elements currently available.

Data on the chemical elements have been the fundamentals of scientific work for years, yet new research is continually revising previously published material about them. Even physical "constants" are subject to change in the light of additional research.

The information contained in this the third edition reflects state-of-the-art values on the most frequently required constants. The material in this current edition was compiled, corrected, and updated over a period of several years, utilizing hundreds of sources. Each value was checked in a minimum of 10 sources to ensure accuracy. A partial listing of the primary reference sources consulted is given at the end of the monographs.

I wish to acknowledge the significant assistance of Wayne Hruden for updating the reported values of the constants and the support given by the Duracell International Inc.

SAMUEL RUBEN
December 1984

Introduction

This handbook contains monographs for each of the 108 known chemical elements, arranged in alphabetical order for rapid reference.

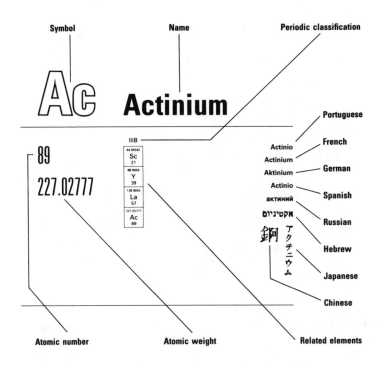

Except where unavailable, values for the following twenty-five different elemental constants are given:

Periodic classification The group, family name, and/or series of the element; this categorization reflects the position of the element in the periodic table.

Atomic number An element of atomic number Z occupies the Zth position in the periodic classification. Its neutral atom has a nucleus with a charge of $+Z\epsilon$ surrounded by Z electrons, each of charge $-\epsilon$.

Atomic weight The relative atomic mass (A_r) based on $^{12}C \equiv 12$; the value for the most stable isotope is given for synthetic elements.

Naturally occurring isotopes   Mass numbers of the isotopes are listed in decreasing order of natural terrestrial abundance.

Density   The weight per unit volume of the element; measurements of this constant are generally made at 25°C, but the temperature utilized is shown in parentheses. Units are grams per cubic centimeter (g/cm^3).

Melting point   Units are degrees Celsius (°C); **Boiling point:**   Units are degrees Celsius (°C).

Latent heat of fusion   The quantity of heat required to change 1 g of the solid element into the liquid state at a constant temperature. Units are Joules per gram (J/g).

Specific heat   The thermal capacity of an element; the specific heat capacity is the quantity of heat required to raise the temperature of a mass through a measured number of Celsius degrees. Units are Joules per gram per degree Celsius (J/g/°C).

Coefficient of lineal thermal expansion   The ratio of the change in length per degree Celsius to the original length at zero degrees Celsius. Units are centimeter per centimeter per degree Celsius (cm/cm/°C).

Thermal conductivity   Thermal energy transmitted through a unit cube per unit time when there exists unit temperature difference between opposite parallel faces. Units are watts (or milliwatts) per centimeter per degree Celsius [w (or mw)/cm/°C].

Electrical resistivity   A proportionality factor (ρ) relating the resistance to current flow between parallel faces of a 1-cm cube of the element. This factor is also known as specific resistance. Because the resistance of semiconductor is substantially influenced by the presence of traces of impurities, the intrinsic resistivity is the parameter given for these ultrapure elements. Units are ohm-centimeters (ohm-cm).

Ionization potential (1st)   The energy necessary to remove the least strongly bound electron from its orbit and place it at rest at an infinite distance. Units are electron volts (eV).

Electron work function (ϕ)   The minimum photonic energy required to remove an electron from the boundary of an element; also known as photoelectric work function. Units are electron volts (eV).

Oxidation potential   The difference in potential produced by a voltaic half-cell associated with the cited chemical reaction. By using the oxidation potential, the likelihood of various chemical reactions can be predicted. Oxidation of gaseous hydrogen (at 1 atmosphere pressure) to ionic hydrogen (in 1 molar acid solution at 25°C) defines the zero reference. Units are volts (V).

Chemical valence   The number of hydrogen atoms (or their equivalent) with which an atom of an element can combine (if negative) or the number which it can displace in a reaction (if positive). The principal valence is set in italic type when more than one valence is possible.

Electrochemical equivalents   The mass of an element displaced by the passage of unit quantity of electricity. The values provided are derived from:

$$\text{electrochemical equivalents} = \frac{kA}{n}$$

where k is a constant equal to 0.0373100, A is the gram-atomic weight, and n is the principal valence. Units are grams per ampere-hour (g/amp-hr).

Ionic radius   The radius an ion exhibits in an ionic crystal in which the ions are packed together with their outermost electronic shells in contact with each other. Values are given for a coordination number of 6. Ionic radii for other coordination numbers can be obtained by multiplying by the following conversion factors:

Coordination Number	Conversion Factor
12	1.12
9	1.05
8	1.03
6	1.00
4	0.94

Units are Ångstroms ($1\text{Å} = 10^{-8}$ cm).

Valence electron potential ($-\epsilon$V)   A calculated value based on the charge of the valence electrons and the ionic radius. It provides a quantitative indication of the reactivity of an element and is determined by the equation:

$$(-\epsilon V) = \frac{kn}{r}$$

where $(-\epsilon V)$ is the valence electron potential, n is the valence, and k is a proportionality factor converting Ångstroms to centimeters and expressing the force exerted by the valence electrons in electron volts and is equal to 14.399; r is the ionic radius in Ångstroms. The principal valence has been used for the determination.

Electronic configuration   A sequential listing of the orbiting electrons, indicating the principal shells and the number of electrons in each subshell. For example, $4d^{10}$ would indicate the presence of 10 electrons in the "d" subshell of the fourth (N) principal shell. Principal shells are assigned letters corresponding to their quantum numbers as follows: 1 = K, 2 = L, 3 = M, 4 = N, 5 = O, 6 = P, and 7 = Q. A maximum exists for the number of electrons in each subshell: 2 in s, 6 in p, 10 in d, and 14 in f.

Valence electrons   A sequential listing of the electrons involved in the ionization of the element. They are indicated in the same manner as in the electronic configuration.

Crystal form   A brief description of the atomic arrangement in the elemental solid state. (See accompanying figure for common Crystal Forms).

Half life The time required for one-half of an initial quantity of a radioactive isotope to be converted into its decay product. This entry is included only when all known isotopes of an element are unstable. The half life presented is that of the most stable isotope. Units are seconds, minutes, hours, days, or years.

Cross section σ The effective size of a nucleus in capturing a thermal (slow) neutron. The larger the cross section the greater is the probability of neutron capture. Units are millibarns (mbarns) or barns (1 barn = 10^{-24} cm^2).

Vapor pressure The pressure exerted when a solid or liquid is in equilibrium with its vapor. Since this parameter is a function of temperature, the vapor pressure at the melting point is given. Units are Pascals (Pa).

Crystal Forms

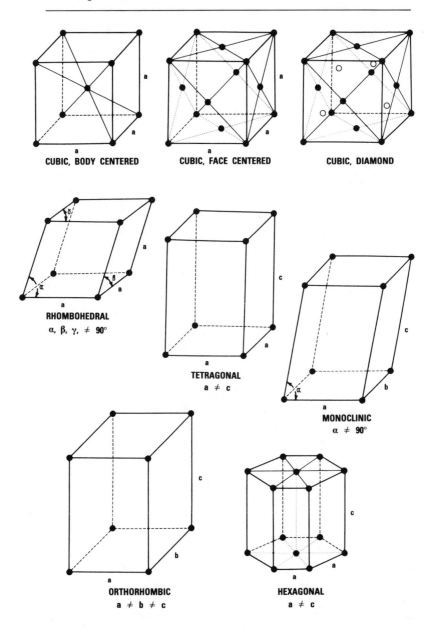

CUBIC, BODY CENTERED

CUBIC, FACE CENTERED

CUBIC, DIAMOND

RHOMBOHEDRAL
α, β, γ, ≠ 90°

TETRAGONAL
a ≠ c

MONOCLINIC
α ≠ 90°

ORTHORHOMBIC
a ≠ b ≠ c

HEXAGONAL
a ≠ c

Periodic Classification of the Elements

IA									
1.0079 **H** 1	IIA								
6.941 **Li** 3	9.01218 **Be** 4								
22.98977 **Na** 11	24.305 **Mg** 12	IIIB	IVB	VB	VIB	VIIB	|——— VIII ———|		
39.098 **K** 19	40.08 **Ca** 20	44.95592 **Sc** 21	47.90 **Ti** 22	50.9415 **V** 23	51.996 **Cr** 24	54.9380 **Mn** 25	55.847 **Fe** 26	58.9332 **Co** 27	58.70 **Ni** 28
85.4678 **Rb** 37	87.62 **Sr** 38	88.9059 **Y** 39	91.22 **Zr** 40	92.9064 **Nb** 41	95.94 **Mo** 42	96.906 **Tc** 43	101.07 **Ru** 44	102.9055 **Rh** 45	106.4 **Pd** 46
132.9054 **Cs** 55	137.34 **Ba** 56	138.9055 **La*** 57	178.49 **Hf** 72	180.9479 **Ta** 73	183.85 **W** 74	186.2 **Re** 75	190.2 **Os** 76	192.22 **Ir** 77	195.09 **Pt** 78
223.01976 **Fr** 87	226.02544 **Ra** 88	227.02777 **Ac†** 89	**104**	**105**	**106**	**107**		**109**	

*Lanthanide Series

140.12 **Ce** 58	140.9077 **Pr** 59	144.24 **Nd** 60	144.913 **Pm** 61	150.4 **Sm** 62	151.96 **Eu** 63	157.25 **Gd** 64	158.9254 **Tb** 65	162.50 **Dy** 66	164.9304 **Ho** 67

†Actinide Series

232.03807 **Th** 90	231.0359 **Pa** 91	238.029 **U** 92	237.0482 **Np** 93	244.06423 **Pu** 94	243.0614 **Am** 95	247.07038 **Cm** 96	247.07032 **Bk** 97	251.07961 **Cf** 98	254.08805 **Es** 99

IB	IIB	IIIA	IVA	VA	VIA	VIIA	O
							4.00260 **He** 2
		10.81 **B** 5	12.011 **C** 6	14.0067 **N** 7	15.9994 **O** 8	18.998403 **F** 9	20.179 **Ne** 10
		26.98154 **Al** 13	28.0855 **Si** 14	30.97376 **P** 15	32.06 **S** 16	35.453 **Cl** 17	39.948 **Ar** 18
63.546 **Cu** 29	65.38 **Zn** 30	69.72 **Ga** 31	72.59 **Ge** 32	74.9216 **As** 33	78.96 **Se** 34	79.904 **Br** 35	83.80 **Kr** 36
107.868 **Ag** 47	112.41 **Cd** 48	114.82 **In** 49	118.69 **Sn** 50	121.75 **Sb** 51	127.60 **Te** 52	126.9045 **I** 53	131.30 **Xe** 54
196.9665 **Au** 79	200.59 **Hg** 80	204.37 **Tl** 81	207.2 **Pb** 82	208.9804 **Bi** 83	208.98243 **Po** 84	209.987 **At** 85	222.01761 **Rn** 86

167.26 **Er** 68	168.9342 **Tm** 69	173.04 **Yb** 70	174.97 **Lu** 71

257.09515 **Fm** 100	258 **Md** 101	259 **No** 102	260 **Lr** 103

Ac Actinium

89

227.02777

IIIB
44.95592 **Sc** 21
88.9059 **Y** 39
138.9055 **La** 57
227.02777 **Ac** 89

Actínio

Actinium

Aktinium

Actinio

актиний

Naturally occurring isotope: 227 (minute quantities only)

Density: 10.07 g/cm³ (25°C)

Melting point: 1100 ± 50°C   **Boiling point:** 3200 ± 300°C (est)

Latent heat of fusion: 62 J/g

Specific heat: 0.12 J/g/°C

Thermal conductivity: 0.12 w/cm/°C (25°C)

Ionization potential (1st): 5.17 eV

Oxidation potential: $Ac \rightarrow Ac^{3+} + 3\epsilon = 2.2$ V

Chemical valence: 3

Electrochemical equivalents: 2.82347 g/amp-hr

Ionic radius: 1.119 Å (Ac^{3+})

Valence electron potential ($-\epsilon V$): 38.60 (Ac^{3+})

Principal quantum number: 7

Principal electron shells: K L M N O P Q

Electronic configuration: $1s^2\ 2s^2\ 2p^6\ 3s^2\ 3p^6\ 3d^{10}\ 4s^2\ 4p^6\ 4d^{10}\ 4f^{14}\ 5s^2\ 5p^6$ $5d^{10}\ 6s^2\ 6p^6\ 6d^1\ 7s^2$

Valence electrons: $6d^1\ 7s^2$

Crystal form: Cubic, face centered

Half life: 21.77 years

Cross section σ: 810 ± 20 barns

Aluminum

13

26.98154

IIIA
10.81 **B** 5
26.98154 **Al** 13
69.72 **Ga** 31
114.82 **In** 49
204.37 **Tl** 81

Alumínio
Aluminium
Aluminium
Aluminio
алюминий
אלומין
鋁 アルミニウム

Naturally occurring isotope: 27
Density: 2.6984 g/cm^3 (20°C)
Melting point: 660.37°C   **Boiling point:** 2467°C
Latent heat of fusion: 395.7 J/g
Specific heat: 0.903 J/g/°C (25°C)
Coefficient of lineal thermal expansion: 23.9 × 10^{6}cm/cm/°C (20°C)
Thermal conductivity: 2.37 w/cm/°C (25°C)
Electrical resistivity: 2.6548 × 10^{-6} ohm-cm (20°C)
Ionization potential (1st): 5.986 eV
Electron work function φ: 4.28 eV
Oxidation potential: $Al \rightarrow Al^{3+} + 3\epsilon = 1.662$ V
Chemical valence: 3
Electrochemical equivalents: 0.33556 g/amp-hr
Ionic radius: 0.535 Å (Al^{3+})
Valence electron potential ($-\epsilon V$): 80.7
Principal quantum number: 3
Principal electron shells: K L M
Electronic configuration: $1s^2 2s^2 2p^6 3s^2 3p^1$
Valence electrons: $3s^2 3p^1$
Crystal form: Cubic, face centered
Cross section σ: 232 ± 3 mbarns
Vapor pressure: 2.42 × 10^{-6} Pa (at melting point)

 **Americium**

95

243.0614

Actinide Series

232.03807	231.0359	238.029	237.0482	244.06423	243.0614	247.07038	247.07032	251.07961	254.08805
Th	Pa	U	Np	Pu	Am	Cm	Bk	Cf	Es
90	91	92	93	94	95	96	97	98	99
257.09515	258	259	260						
Fm	Md	No	Lr						
100	101	102	103						

Amerício
Américium
Amerizium
Americio
америций
אמריציום

鎇 ア
メ
リ
シ
ウ
ム

Naturally occurring isotopes: None
Density: 13.67 g/cm^3 (20°C)
Melting point: 1176°C   **Boiling point:** 2011°C
Ionization potential (1st): 5.99 eV
Oxidation potential: Am → Am^{3+} + 3ϵ = 2.32 V
Chemical valence: 2, *3*, 4, 5, 6
Electrochemical equivalents: 3.0229 g/amp-hr
Ionic radius: 0.982 Å (Am^{3+})
Valence electron potential ($-\epsilon$V): 44.0
Principal quantum number: 7
Principal electron shells: K L M N O P Q
Electronic configuration: 1s^2 2s^2 2p^6 3s^2 3p^6 3d^{10} 4s^2 4p^6 4d^{10} 4f^{14} 5s^2 5p^6
    5d^{10} 5f^7 6s^2 6p^6 7s^2
Valence electrons: 5f^7 7s^2
Crystal form: Hexagonal
Half life: 7.32 × 10^3 years
Cross section σ: 180 ± 20 barns

Antimony

51

121.75

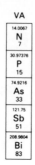

VA

| 14.0067 |
| N |
| 7 |
| 30.97376 |
| P |
| 15 |
| 74.9216 |
| As |
| 33 |
| 121.75 |
| Sb |
| 51 |
| 208.9804 |
| Bi |
| 83 |

Antimônio

Antimoine

Antimon

Antimonio

сурьма

אנטימון

銻 アソチモソ

Naturally occurring isotopes: 121, 123
Density: 6.691 g/cm^3 (20°C)
Melting point: 630.74°C   **Boiling point:** 1750°C
Latent heat of fusion: 165.0 J/g
Specific heat: 0.207 J/g/°C (25°C)
Coefficient of lineal thermal expansion: 9.2 × 10^{-6} cm/cm/°C (0°C)
Thermal conductivity: 0.244 w/cm/°C (25°C)
Electrical resistivity: 39 × 10^{-6} ohm-cm (0°C)
Ionization potential (1st): 8.641 eV
Electron work function φ: 4.55 eV
Oxidation potential: $2Sb + 3H_2O \rightarrow Sb_2O_3 + 6H^+ + 6\epsilon = -0.152$ V
Chemical valence: −3, 0, *3*, 5
Electrochemical equivalents: 1.5142 g/amp-hr
Ionic radius: 0.76 Å (Sb^{3+})
Valence electron potential (−εV): 57
Principal quantum number: 5
Principal electron shells: K L M N O
Electronic configuration: $1s^2\ 2s^2\ 2p^6\ 3s^2\ 3p^6\ 3d^{10}\ 4s^2\ 4p^6\ 4d^{10}\ 5s^2\ 5p^3$
Valence electrons: $5s^2\ 5p^3$
Crystal form: Rhombohedral
Cross section σ: 5 ± 1 barns
Vapor pressure: 2.49 × 10^{-9} Pa (at melting point)

Argon

18

39.948

O
4.00260 He 2
20.179 Ne 10
39.948 Ar 18
83.80 Kr 36
131.30 Xe 54
222.01761 Rn 86

Argônio

Argon

Argon

Argón

аргон

ארגון

氬 アルゴン

Naturally occurring isotopes: 40, 36, 38
Density: 1.65 g/cm^3 ($-233°C$), 1.784 × 10^{-3} g/cm^3 (0°C)
Melting point: $-189.2°C$ **Boiling point:** $-185.7°C$
Latent heat of fusion: 29.45 J/g
Specific heat: 0.52032 J/g/°C (25°C)
Thermal conductivity: 0.1772 mw/cm/°C (27°C at 1 atm)
Ionization potential (1st): 15.759 eV
Chemical valence: 0
Principal quantum number: 3
Principal electron shells: K L M
Electronic configuration: 1s^2 2s^2 2p^6 3s^2 3p^6
Valence electrons: (3s^2 3p^6)
Crystal form: Cubic, face centered
Cross section σ: 0.66 barns

Arsenic

33

74.9216

VA

14.0067	
N	
7	
30.97376	
P	
15	
74.9216	
As	
33	
121.75	
Sb	
51	
208.9804	
Bi	
83	

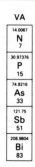

Arsênio

Arsenic

Arsen

Arsénico

мышьяк

ארסן

砷 砒
磇

Naturally occurring isotope: 75
Density: 5.73 g/cm³ (gray) (20°C)
Melting point: 817°C (at 28 atm) **Boiling point:** 613°C (sublimes)
Latent heat of fusion: 369.9 J/g
Specific heat: 0.329 J/g/°C (gray) (25°C)
Coefficient of lineal thermal expansion: 6.02×10^{-6} cm/cm/°C (25°C)
Thermal conductivity: 0.502 w/cm/°C (gray) (25°C)
Electrical resistivity: 35×10^{-6} ohm-cm (0°C)
Ionization potential (1st): 9.81 eV
Electron work function ϕ: 3.75 eV
Oxidation potential: $As + 2H_2O \rightarrow HAsO_2 + 3H^+ + 3\epsilon = -0.2476$ V
Chemical valence: $-3, 0, 3, 5$
Electrochemical equivalents: 0.93177 g/amp-hr
Ionic radius: 0.58 Å (As^{3+})
Valence electron potential $(-\epsilon V)$: 74
Principal quantum number: 4
Principal electron shells: K L M N
Electronic configuration: $1s^2\ 2s^2\ 2p^6\ 3s^2\ 3p^6\ 3d^{10}\ 4s^2\ 4p^3$
Valence electrons: $4s^2\ 4p^3$
Crystal form: Rhombohedral
Cross section σ: 4.30 ± 0.10 barns

Astatine

85

209.987

VIIA

18.998403 F 9	
35.453 Cl 17	
79.904 Br 35	
126.9045 I 53	
209.987 At 85	

Astato

Astatine

Astat

Astatino

астатин

רצצצא

砤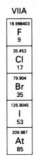

Naturally occurring isotopes: None

Melting point: 302°C (est)   **Boiling point:** 337°C (est)

Latent heat of fusion: 114 J/g (est)

Ionization potential (1st): 9.65 eV

Oxidation potential: $2At^- \rightarrow At_2 + 2\epsilon = -0.2$ V

Chemical valence: *1*, 3, 5, 7

Electrochemical equivalents: 7.8346 g/amp-hr

Principal quantum number: 6

Principal electron shells: K L M N O P

Electronic configuration: $1s^2\ 2s^2\ 2p^6\ 3s^2\ 3p^6\ 3d^{10}\ 4s^2\ 4p^6\ 4d^{10}\ 4f^{14}\ 5s^2\ 5p^6$ $5d^{10}\ 6s^2\ 6p^5$

Valence electrons: $6s^2\ 6p^5$

Half life: 8.1 hr

Barium

56

137.34

IIA
9.01218 **Be** 4
24.305 **Mg** 12
40.08 **Ca** 20
87.62 **Sr** 38
137.34 **Ba** 56
226.02544 **Ra** 88

Bário

Barium

Barium

Bario

барий

בריום

鋇 バリウム

Naturally occurring isotopes: 138, 137, 136, 135, 134, 130, 132
Density: 3.59 g/cm^3 (20°C)
Melting point: 725°C   **Boiling point:** 1640°C
Latent heat of fusion: 55.79 J/g
Specific heat: 0.204 J/g/°C (25°C)
Coefficient of lineal thermal expansion: 19.0 × 10^{-6} cm/cm/°C (20°C)
Thermal conductivity: 0.184 w/cm/°C (22°C)
Ionization potential (1st): 5.212 eV
Electron work function φ: 2.7 eV
Oxidation potential: Ba → Ba^{2+} + 2ε = 2.906 V
Chemical valence: 2
Electrochemical equivalents: 2.5621 g/amp-hr
Ionic radius: 1.35 Å (Ba^{2+})
Valence electron potential (−εV): 21.3
Principal quantum number: 6
Principal electron shells: K L M N O P
Electronic configuration: 1s^2 2s^2 2p^6 3s^2 3p^6 3d^{10} 4s^2 4p^6 4d^{10} 5s^2 5p^6 6s^2
Valence electrons: 6s^2
Crystal form: Cubic, body centered
Cross section σ: 1.2 ± 0.1 barns
Vapor pressure: 9.80 × 10 Pa (at melting point)

Berkelium

97

247.07032

Actinide Series

232.03807	231.0359	238.029	237.0482	244.06423	243.0614	247.07038	247.07032	251.07961	254.08805
Th	Pa	U	Np	Pu	Am	Cm	Bk	Cf	Es
90	91	92	93	94	95	96	97	98	99

257.09515	258	259	260
Fm	Md	No	Lr
100	101	102	103

Berquélio
Berkelium
Berkelium
Berkelio
беркелий
ברקליום
錇 バークリウム

Naturally occurring isotopes: None
Density: 14.78 g/cm^3 (25°C)
Melting point: 986 ± 25°C
Ionization potential (1st): 6.23 eV
Oxidation potential: Bk → Bk^{3+} + 3ϵ = 1.97 V
Chemical valence: *3*, 4
Electrochemical equivalents: 3.0727 g/amp-hr
Ionic radius: 0.949 Å (Bk^{3+})
Valence electron potential ($-\epsilon$V): 45.5
Principal quantum number: 7
Principal electron shells: K L M N O P Q
Electronic configuration: 1s^2 2s^2 2p^6 3s^2 3p^6 3d^{10} 4s^2 4p^6 4d^{10} 4f^{14} 5s^2 5p^6
5d^{10} 5f^8 6s^2 6p^6 6d^1 7s^2
Valence electrons: 5f^8 6d^1 7s^2
Crystal form: Hexagonal
Half life: 1.4 × 10^3 years

Beryllium

4

9.01218

IIA

9.01218	Be 4
24.305	Mg 12
40.08	Ca 20
87.62	Sr 38
137.34	Ba 56
226.02544	Ra 88

Berílio
Beryllium
Beryllium
Berilio
бериллий
בר- ליום

鈹 バリリウム

Naturally occurring isotope: 9
Density: 1.848 g/cm^3 (20°C)
Melting point: 1278±5°C   **Boiling point:** 2970°C
Latent heat of fusion: 1301 J/g
Specific heat: 1.82 J/g/°C (25°C)
Coefficient of lineal thermal expansion: 11.6 × 10^{-6} cm/cm/°C (20°C)
Thermal conductivity: 2.01 w/cm/°C (25°C)
Electrical resistivity: 4.0 × 10^{-6} ohm-cm (20°C)
Ionization potential (1st): 9.322 eV
Electron work function ϕ: 4.98 eV
Oxidation potential: Be → Be^{2+} + 2ϵ = 1.85 V
Chemical valence: 2
Electrochemical equivalents: 0.16812 g/amp-hr
Ionic radius: 0.35 Å (Be^{2+})
Valence electron potential ($-\epsilon$V): 82
Principal quantum number: 2
Principal electron shells: K L
Electronic configuration: 1s^2 2s^2
Valence electrons: 2s^2
Crystal form: Hexagonal, close packed
Cross section σ: 9.2±0.5 mbarns
Vapor pressure: 4.18 Pa (at melting point)

Bismuth

83

208.9804

VA
14.0067 N 7
30.97376 P 15
74.9216 As 33
121.75 Sb 51
208.9804 Bi 83

Bismuto

Bismuth

Wismut

Bismuto

висмут

ביזמוט

鉍 ビスマス

Naturally occurring isotope: 209
Density: 9.78 g/cm³ (20°C)
Melting point: 271.3°C   **Boiling point:** 1560 ± 5°C
Latent heat of fusion: 52.09 J/g
Specific heat: 0.122 J/g/°C (25°C)
Coefficient of lineal thermal expansion: 13.3×10^{-6} cm/cm/°C
Thermal conductivity: 0.0792 w/cm/°C (25°C)
Electrical resistivity: 106.8×10^{-6} ohm-cm (0°C)
Ionization potential (1st): 7.289 eV
Electron work function φ: 4.22 eV
Oxidation potential: $Bi + H_2O \rightarrow BiO^+ + 2H^+ + 3\epsilon = -0.320$ V
Chemical valence: *3*, 5
Electrochemical equivalents: 2.5990 g/amp-hr
Ionic radius: 1.03 Å (Bi^{3+})
Valence electron potential ($-\epsilon$V): 41.9
Principal quantum number: 6
Principal electron shells: K L M N O P
Electronic configuration: $1s^2\ 2s^2\ 2p^6\ 3s^2\ 3p^6\ 3d^{10}\ 4s^2\ 4p^6\ 4d^{10}\ 4f^{14}\ 5s^2\ 5p^6$
    $5d^{10}\ 6s^2\ 6p^3$
Valence electrons: $6s^2\ 6p^3$
Crystal form: Rhombohedral
Cross section σ: 19 ± 2 mbarns
Vapor pressure: 6.27×10^{-4} Pa (at melting point)

Boron

5

10.81

| 10.81
B
5 |
| 26.98154
Al
13 |
| 69.72
Ga
31 |
| 114.82
In
49 |
| 204.37
Tl
81 |

Bóro

Bore

Bor

Boro

бор

בור

硼 素

Naturally occurring isotopes: 11, 10
Density: 2.34 g/cm^3 (crystalline), 2.37 g/cm^3 (amorphous) (both at 20°C)
Melting point: 2300°C   **Boiling point:** 2550°C (sublimes)
Latent heat of fusion: 890.8 J/g
Specific heat: 1.03 J/g/°C (25°C)
Coefficient of lineal thermal expansion: 8.3 × 10^{-6} cm/cm/°C (20°C)
Thermal conductivity: 0.274 w/cm/°C (25°C)
Electrical resistivity: 1.8 × 10^6 ohm-cm (0°C)
Ionization potential (1st): 8.298 eV
Electron work function ϕ: 4.45 eV
Oxidation potential: B + 3H$_2$O → H$_3$BO$_3$ + 3H$^+$ + 3ϵ = −0.8698 V
Chemical valence: 3
Electrochemical equivalents: 0.1344 g/amp-hr
Ionic radius: 0.23 Å (B^{3+})
Valence electron potential (−ϵV): 190
Principal quantum number: 2
Principal electron shells: K L
Electronic configuration: 1s^2 2s^2 2p^1
Valence electrons: 2s^2 2p^1
Crystal form: Hexagonal, close packed
Cross section σ: 759 barns
Vapor pressure: 3.48 × 10^{-1} Pa (at melting point)

Br Bromine

35

79.904

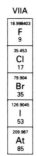

VIIA
18.998403 F 9
35.453 Cl 17
79.904 Br 35
126.9045 I 53
209.987 At 85

Bromo
Brome
Brom
Bromo
бром
ברום

溴 臭素

Naturally occurring isotopes: 79, 81
Density: 3.1028 g/cm^3 (20°C)
Melting point: −7.2°C **Boiling point:** 58.78°C
Latent heat of fusion: 132.0 J/g (Br$_2$)
Specific heat: 0.47362 J/g/°C (Br$_2$) (25°C)
Thermal conductivity: 1.22 mw/cm/°C (27°C)
Electrical resistivity: 7.8 × 10^{12} ohm-cm (0°C)
Ionization potential (1st): 11.814 eV
Oxidation potential: $2Br^- \rightarrow Br_2 + 2\epsilon = -1.0652$ V
Chemical valence: − 1, 3, 5, 7
Electrochemical equivalents: 2.9812 g/amp-hr
Ionic radius: 1.96 Å (Br$^-$)
Valence electron potential (−εV): −7.35
Principal quantum number: 4
Principal electron shells: K L M N
Electronic configuration: $1s^2\ 2s^2\ 2p^6\ 3s^2\ 3p^6\ 3d^{10}\ 4s^2\ 4p^5$
Valence electrons: $4s^2\ 4p^5$
Crystal form: Orthorhombic, rhombic
Cross section σ: 6.8 ± 0.1 barns
Vapor pressure: 5.80 × 10^3 Pa (at melting point)

Cadmium

48

112.41

	IIB
	65.38
	Zn
	30
	112.41
	Cd
	48
	200.59
	Hg
	80

Cádmio
Cadmium
Cadmium
Cadmio
кадмий
קדמיום

カドミウム

Naturally occurring isotopes: 114, 112, 111, 110, 113, 116, 106, 108
Density: 8.65 g/cm^3 (20°C)
Melting point: 320.9°C   **Boiling point:** 765°C
Latent heat of fusion: 54.01 J/g
Specific heat: 0.231 J/g/°C (25°C)
Coefficient of lineal thermal expansion: 29.8 × 10^{-6} cm/cm/°C (25°C)
Thermal conductivity: 0.969 w/cm/°C (25°C)
Electrical resistivity: 6.83 × 10^{-6} ohm-cm (0°C)
Ionization potential (1st): 8.993 eV
Electron work function φ: 4.22 eV
Oxidation potential: $Cd \rightarrow Cd^{2+} + 2\epsilon = 0.4029$ V
Chemical valence: 2
Electrochemical equivalents: 2.0970 g/amp-hr
Ionic radius: 0.97 Å (Cd^{2+})
Valence electron potential ($-\epsilon$V): 30
Principal quantum number: 5
Principal electron shells: K L M N O
Electronic configuration: $1s^2\ 2s^2\ 2p^6\ 3s^2\ 3p^6\ 3d^{10}\ 4s^2\ 4p^6\ 4d^{10}\ 5s^2$
Valence electrons: $5s^2$
Crystal form: Hexagonal, close packed
Cross section σ: 2450 ± 20 barns
Vapor pressure: 1.48 × 10 Pa (at melting point)

Calcium

IIA		
9.01218 Be 4		
24.305 Mg 12		
40.08 Ca 20		
87.62 Sr 38		
137.34 Ba 56		
226.02544 Ra 88		

20

40.08

Cálcio

Calcium

Kalzium

Calcio

кальций

סידן

鈣 カルシウム

Naturally occurring isotopes: 40, 44, 42, 48, 43, 46
Density: 1.55 g/cm^3 (20°C)
Melting point: 839 ± 2°C   **Boiling point:** 1484°C
Latent heat of fusion: 216.2 J/g
Specific heat: 0.632 J/g/°C (25°C)
Coefficient of lineal thermal expansion: 22.3 × 10^{-6} cm/cm/°C (20°C)
Thermal conductivity: 2.01 w/cm/°C (25°C)
Electrical resistivity: 3.91 × 10^{-6} ohm-cm (0°C)
Ionization potential (1st): 6.113 eV
Electron work function ϕ: 2.87 eV
Oxidation potential: $Ca \rightarrow Ca^{2+} + 2\epsilon = 2.866$ V
Chemical valence: 2
Electrochemical equivalents: 0.7477 g/amp-hr
Ionic radius: 0.99 Å (Ca^{2+})
Valence electron potential ($-\epsilon V$): 29
Principal quantum number: 4
Principal electron shells: K L M N
Electronic configuration: $1s^2\ 2s^2\ 2p^6\ 3s^2\ 3p^6\ 4s^2$
Valence electrons: $4s^2$
Crystal form: Cubic, face centered
Cross section σ: 0.44 ± 0.02 barns
Vapor pressure: 2.54 × 10^2 Pa (at melting point)

Californium

98

251.07961

Actinide Series

232.03807	231.0359	238.029	237.0482	244.06423	243.0614	247.07038	247.07032	251.07961	254.08805
Th	Pa	U	Np	Pu	Am	Cm	Bk	Cf	Es
90	91	92	93	94	95	96	97	98	99

257.09515	258	259	260
Fm	Md	No	Lr
100	101	102	103

Califórnio
Californium
Californium
Californio
калифорний

קליפורניום

鋼 ㊙ カリフォリニウム

Naturally occurring isotopes: None
Density: 15.1 g/cm^3 (25°C)
Melting point: 900 ± 30°C
Ionization potential (1st): 6.30 eV
Oxidation potential: Cf → Cf^{3+} + 3ϵ = 2.0 V
Chemical valence: 2, *3*, 4
Electrochemical equivalents: 3.1226 g/amp-hr
Ionic radius: 0.934 Å (Cf^{3+})
Valence electron potential ($-\epsilon$V): 44.5
Principal quantum number: 7
Principal electron shells: K L M N O P Q
Electronic configuration: 1s^2 2s^2 2p^6 3s^2 3p^6 3d^{10} 4s^2 4p^6 4d^{10} 4f^{14} 5s^2 5p^6
    5d^{10} 5f^{10} 6s^2 6p^6 7s^2
Valence electrons: 5f^{10} 7s^2
Crystal form: Hexagonal
Half life: 900 years
Cross section σ: 2100 ± 1000 barns

Carbon

6	IVA
12.011	

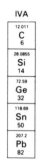

12.011	
C	
6	
28.0855	
Si	
14	
72.59	
Ge	
32	
118.69	
Sn	
50	
207.2	
Pb	
82	

Carbono

Carbone

Kohlenstoff

Carbono

углерод

פחמן

碳 炭素

Naturally occurring isotopes: 12, 13, 14

Density: 3.52 g/cm^3 (diamond), 1.9–2.3 g/cm^3 (graphite), 1.8–2.1 g/cm^3 (amorphous) (all at 20°C)

Melting point: 3550°C   **Boiling point:** 4827°C

Specific heat: 0.7099 J/g/°C (graphite) (25°C)

Coefficient of lineal thermal expansion: 2.10 × 10^{-6} cm/cm/°C (graphite) (30°C)

Thermal conductivity: 0.8–2.2 w/cm/°C (graphite) (25°C)

Electrical resistivity: 1375 × 10^{-6} ohm-cm (graphite) (0°C)

Ionization potential (1st): 11.260 eV

Electron work function ϕ: 5.0 eV

Oxidation potential: $CH_4 \rightarrow C + 4H^+ + 4\epsilon = -0.1316$ V

Chemical valence: 2, 3, *4*

Electrochemical equivalents: 0.11203 g/amp-hr

Ionic radius: 0.16 Å (C^{4+})

Valence electron potential $(-\epsilon V)$: 360

Principal quantum number: 2

Principal electron shells: K L

Electronic configuration: $1s^2\ 2s^2\ 2p^2$

Valence electrons: $2s^2\ 2p^2$

Crystal form: Hexagonal (graphite), cubic (diamond)

Cross section σ: 3.4±0.2 mbarns

Ce Cerium

58

140.12

Lanthanide Series

Cério
Cérium
Zerium
Cerio
церий
צריום

鈰 セリウム

140.12	140.9077	144.24	144.913	150.4	151.96	157.25	158.9254	162.50	164.9304
Ce	Pr	Nd	Pm	Sm	Eu	Gd	Tb	Dy	Ho
58	59	60	61	62	63	64	65	66	67
167.26	168.9342	173.04	174.97						
Er	Tm	Yb	Lu						
68	69	70	71						

Naturally occurring isotopes: 140, 142, 138, 136
Density: 6.657 g/cm^3 (25°C)
Melting point: 799°C   **Boiling point:** 3426°C
Latent heat of fusion: 65.7 J/g
Specific heat: 0.192 J/g/°C (25°C)
Coefficient of lineal thermal expansion: 7.1 × 10^{-6} cm/cm/°C (25°C)
Thermal conductivity: 0.113 w/cm/°C (25°C)
Electrical resistivity: 77 × 10^{-6} ohm-cm (25°C)
Ionization potential (1st): 5.47 eV
Electron work function ϕ: 2.84 eV
Oxidation potential: $Ce \rightarrow Ce^{3+} + 3\epsilon = 2.483$ V
Chemical valence: *3*, 4
Electrochemical equivalents: 1.7426 g/amp-hr
Ionic radius: 1.034 Å (Ce^{3+})
Valence electron potential ($-\epsilon V$): 41.78
Principal quantum number: 6
Principal electron shells: K L M N O P
Electronic configuration: $1s^2\ 2s^2\ 2p^6\ 3s^2\ 3p^6\ 3d^{10}\ 4s^2\ 4p^6\ 4d^{10}\ 4f^2\ 5s^2\ 5p^6\ 6s^2$
Valence electrons: $4f^2\ 6s^2$
Crystal form: Cubic, face centered
Cross section σ: 0.73±0.08 barns

Cs Cesium

55

132.9054

IA
1.0079 **H** 1
6.941 **Li** 3
22.98977 **Na** 11
39.098 **K** 19
85.4678 **Rb** 37
132.9054 **Cs** 55
223.01976 **Fr** 87

Césio
Césium
Caesium
Ceslo
цезий
צזיום
鉇 セシウム

Naturally occurring isotope: 133
Density: 1.873 g/cm^3 (20°C)
Melting point: 28.40 ± 0.01°C   **Boiling point:** 669.3°C
Latent heat of fusion: 16.372 J/g
Specific heat: 0.241 J/g/°C (25°C)
Coefficient of lineal thermal expansion: 97 × 10^{-6} cm/cm/°C (20°C)
Thermal conductivity: 0.359 w/cm/°C (solid at melting point)
Electrical resistivity: 20.46 × 10^{-6} ohm-cm (20°C)
Ionization potential (1st): 3.894 eV
Electron work function ϕ: 2.14 eV
Oxidation potential: $Cs \rightarrow Cs^+ + \epsilon = 2.923$ V
Chemical valence: 1
Electrochemical equivalents: 4.95870 g/amp-hr
Ionic radius: 1.67 Å (Cs$^+$)
Valence electron potential ($-\epsilon$V): 8.62
Principal quantum number: 6
Principal electron shells: K L M N O P
Electronic configuration: $1s^2\ 2s^2\ 2p^6\ 3s^2\ 3p^6\ 3d^{10}\ 4s^2\ 4p^6\ 4d^{10}\ 5s^2\ 5p^6\ 6s^1$
Valence electrons: 6s^1
Crystal form: Cubic, body centered
Cross section σ: 30.0 ± 1.5 barns
Vapor pressure: 2.50 × 10^{-5} Pa (at melting point)

Chlorine

17

35.453

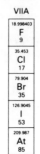

VIIA
18.998403 F 9
35.453 Cl 17
79.904 Br 35
126.9045 I 53
209.987 At 85

Clóro

Chlore

Chlor

Cloro

хлор

כלור

氯 塩素

Naturally occurring isotopes: 35, 37
Density: 1.56 g/cm^3 ($-33.6°C$), 3.214×10^{-3} g/cm^3 (0°C)
Melting point: $-100.98°C$ **Boiling point:** $-34.6°C$
Latent heat of fusion: 180.8 J/g (Cl$_2$)
Specific heat: 0.4782 J/g/°C (Cl$_2$) (25°C)
Thermal conductivity: 0.089 mw/cm/°C (27°C at 1 atm)
Ionization potential (1st): 12.967 eV
Oxidation potential: $2Cl^- \rightarrow Cl_2 + 2\epsilon = -1.3595$ V
Chemical valence: $-I$, 3, 5, 7
Electrochemical equivalents: 1.3228 g/amp-hr
Ionic radius: 1.81 Å (Cl$^-$)
Valence electron potential ($-\epsilon V$): -7.96
Principal quantum number: 3
Principal electron shells: K L M
Electronic configuration: 1s^2 2s^2 2p^6 3s^2 3p^5
Valence electrons: 3s^2 3p^5
Crystal form: Tetragonal
Cross section σ: 33 barns
Vapor pressure: 1.30×10^3 Pa (at melting point)

Chromium

51.996	
Cr	
24	
95.94	
Mo	
42	
183.85	
W	
74	
106	

24

51.996

Crômio
Chrom
Chrom
Cromo
хром
כרום
銘 クローム

Naturally occurring isotopes: 52, 53, 50, 54
Density: 7.20 g/cm^3 (20°C)
Melting point: 1857 ± 20°C   **Boiling point:** 2672°C
Latent heat of fusion: 265.7 J/g
Specific heat: 0.449 J/g/°C (25°C)
Coefficient of lineal thermal expansion: 6.2 × 10^{-6} cm/cm/°C (20°C)
Thermal conductivity: 0.939 w/cm/°C (25°C)
Electrical resistivity: 12.9 × 10^{-6} ohm-cm (20°C)
Ionization potential (1st): 6.766 eV
Electron work function ϕ: 4.5 eV
Oxidation potential: $Cr \rightarrow Cr^{3+} + 3\epsilon = 0.744$ V
Chemical valence: 1, 2, 3, 4, 5, *6*
Electrochemical equivalents: 0.32333 g/amp-hr
Ionic radius: 0.52 Å (Cr^{6+})
Valence electron potential ($-\epsilon$V): 170
Principal quantum number: 4
Principal electron shells: K L M N
Electronic configuration: $1s^2\ 2s^2\ 2p^6\ 3s^2\ 3p^6\ 3d^5\ 4s^1$
Valence electrons: $3d^5\ 4s^1$
Crystal form: Cubic, body centered
Cross section σ: 3.1 ± 0.2 barns
Vapor pressure: 9.90 × 10^2 Pa (at melting point)

Cobalt

27

58.9332

VIII		
55.847 Fe 26	58.9332 Co 27	58.70 Ni 28
101.07 Ru 44	102.9055 Rh 45	106.4 Pd 46
190.2 Os 76	192.22 Ir 77	195.09 Pt 78
	109	

Cobalto

Cobalt

Kobalt

Cobalto

кобальт

קובלט

鈷 コバルト

Naturally occurring isotope: 59
Density: 8.71 g/cm³ (21°C)
Melting point: 1495°C   **Boiling point:** 2870°C
Latent heat of fusion: 258.6 J/g
Specific heat: 4.21 J/g/°C (25°C)
Coefficient of lineal thermal expansion: 13.80×10^{-6} cm/cm/°C (20°C)
Thermal conductivity: 1.00 w/cm/°C (25°C)
Electrical resistivity: 6.24×10^{-6} ohm-cm (20°C)
Ionization potential (1st): 7.86 eV
Electron work function φ: 5.0 eV
Oxidation potential: $Co \rightarrow Co^{2+} + 2\epsilon = 0.277$ V
Chemical valence: *2*, 3, 4
Electrochemical equivalents: 1.0994 g/amp-hr
Ionic radius: 0.745 Å (Co^{3+})
Valence electron potential ($-\epsilon V$): 38.7
Principal quantum number: 4
Principal electron shells: K L M N
Electronic configuration: $1s^2\ 2s^2\ 2p^6\ 3s^2\ 3p^6\ 3d^7\ 4s^2$
Valence electrons: $3d^7\ 4s^2$
Crystal form: Hexagonal, close packed
Cross section σ: 37.5 ± 0.2 barns
Vapor pressure: 1.75×10^2 Pa (at melting point)

Cu — Copper

IB	
63.546 **Cu** 29	
107.868 **Ag** 47	
196.9665 **Au** 79	

Cobre
Cuivre
Kupfer
Cobre
медь

נחושת

銅 銅

Naturally occurring isotopes: 63, 65
Density: 8.96 g/cm³ (25°C)
Melting point: 1083.4 ± 0.2°C   **Boiling point:** 2567°C
Latent heat of fusion: 205.6 J/g
Specific heat: 0.3845 J/g/°C (25°C)
Coefficient of lineal thermal expansion: 16.6×10^{-6} cm/cm/°C (25°C)
Thermal conductivity: 4.01 w/cm/°C (25°C)
Electrical resistivity: 1.678×10^{-6} ohm-cm (20°C)
Ionization potential (1st): 7.726 eV
Electron work function ϕ: 4.65 eV
Oxidation potentials: $Cu \rightarrow Cu^{+} + \epsilon = -0.521$ V
$Cu \rightarrow Cu^{2+} + 2\epsilon = -0.3419$ V
Chemical valence: 1, *2*
Electrochemical equivalents: 1.1855 g/amp-hr
Ionic radius: 0.73 Å (Cu^{2+})
Valence electron potential ($-\epsilon$V): 34
Principal quantum number: 4
Principal electron shells: K L M N
Electronic configuration: $1s^2\ 2s^2\ 2p^6\ 3s^2\ 3p^6\ 3d^{10}\ 4s^1$
Valence electrons: $3d^{10}\ 4s^1$
Crystal form: Cubic, face centered
Cross section σ: 3.8 ± 0.1 barns
Vapor pressure: 5.05×10^{-2} Pa (at melting point)

 **Curium**

96

247.07038

Cúrio
Curium
Curium
Curio
кюрий
קיוריום

鋦 キュリウム

Actinide Series

| 232.03807 Th 90 | 231.0359 Pa 91 | 238.029 U 92 | 237.0482 Np 93 | 244.06423 Pu 94 | 243.0614 Am 95 | 247.07038 Cm 96 | 247.07032 Bk 97 | 251.07961 Cf 98 | 254.08805 Es 99 |
| 257.09515 Fm 100 | 258 Md 101 | 259 No 102 | 260 Lr 103 | | | | | | |

Naturally occurring isotopes: None

Density: 13.51 g/cm^3 (25°C)

Melting point: 1340±40°C   **Boiling point:** 3110°C

Ionization potential (1st): 6.02 eV

Oxidation potential: Cm → Cm^{3+} + 3ϵ = 2.07 V

Chemical valence: *3,* 4

Electrochemical equivalents: 3.0727 g/amp-hr

Ionic radius: 0.970 Å (Cm^{3+})

Valence electron potential ($-\epsilon$V): 44.5

Principal quantum number: 7

Principal electron shells: K L M N O P Q

Electronic configuration: 1s^2 2s^2 2p^6 3s^2 3p^6 3d^{10} 4s^2 4p^6 4d^{10} 4f^{14} 5s^2 5p^6 5d^{10} 5f^7 6s^2 6p^6 6d^1 7s^2

Valence electrons: 5f^7 6d^1 7s^2

Crystal form: Hexagonal

Half life: 1.6 × 10^7 years

Cross section σ: 180 barns

Dy Dysprosium

66

162.50

Lanthanide Series

140.12 Ce 58	140.9077 Pr 59	144.24 Nd 60	144.913 Pm 61	150.4 Sm 62	151.96 Eu 63	157.25 Gd 64	158.9254 Tb 65	162.50 Dy 66	164.9304 Ho 67
167.26 Er 68	168.9342 Tm 69	173.04 Yb 70	174.97 Lu 71						

Dispósio
Dysprosium
Dysprosium
Disprosio
диспрозий
דיספרוזיום

 ジスプロシウム

Naturally occurring isotopes: 164, 162, 163, 161, 160, 158, 156
Density: 8.550 g/cm^3 (25°C)
Melting point: 1412°C **Boiling point:** 2562°C
Latent heat of fusion: 105.6 J/g
Specific heat: 173 J/g/°C (25°C)
Coefficient of lineal thermal expansion: 8.6 × 10^{-6} cm/cm/°C (25°C)
Thermal conductivity: 0.107 w/cm/°C (25°C)
Electrical resistivity: 90 × 10^{-6} ohm-cm (25°C)
Ionization potential (1st): 5.928 eV
Oxidation potential: Dy → Dy^{3+} + 3ε = 2.353 V
Chemical valence: 3
Electrochemical equivalents: 2.0210 g/amp-hr
Ionic radius: 0.912 Å (Dy^{3+})
Valence electron potential ($-$εV): 47.4
Principal quantum number: 6
Principal electron shells: K L M N O P
Electronic configuration: 1s^2 2s^2 2p^6 3s^2 3p^6 3d^{10} 4s^2 4p^6 4d^{10} 4f^{10} 5s^2
5p^6 6s^2
Valence electrons: 4f^{10} 6s^2
Crystal form: Hexagonal, close packed
Cross section σ: 930 ± 20 barns

Es Einsteinium

99

254.08805

Actinide Series

232.03807	231.0359	238.029	237.0482	244.06423	243.0614	247.07038	247.07032	251.07961	254.08805
Th	Pa	U	Np	Pu	Am	Cm	Bk	Cf	Es
90	91	92	93	94	95	96	97	98	99
257.09515	258	259	260						
Fm	Md	No	Lr						
100	101	102	103						

Einstênio
Einsteinium
Einsteinium
Einstenio
эйнштейний
איינסטיניום

Naturally occurring isotopes: None
Melting point: 860 ± 30°C
Ionization potential (1st): 6.42 eV
Oxidation potential: $Es \rightarrow Es^{2+} + 2\epsilon = 2.3$ V
Chemical valence: *2*, 3
Electrochemical equivalents: 4.7400 g/amp-hr
Ionic radius: 0.925 Å (Es^{3+})
Principal quantum number: 7
Principal electron shells: K L M N O P Q
Electronic configuration: $1s^2\ 2s^2\ 2p^6\ 3s^2\ 3p^6\ 3d^{10}\ 4s^2\ 4p^6\ 4d^{10}\ 4f^{14}\ 5s^2\ 5p^6$
      $5d^{10}\ 5f^{11}\ 6s^2\ 6p^6\ 7s^2$
Valence electrons: $5f^{11}\ 7s^2$
Crystal form: Cubic, face centered
Half life: 276 days
Cross section σ: < 40 barns

Erbium

68

167.26

Lanthanide Series

140.12	140.9077	144.24	144.913	150.4	151.96	157.25	158.9254	162.50	164.9304
Ce	Pr	Nd	Pm	Sm	Eu	Gd	Tb	Dy	Ho
58	59	60	61	62	63	64	65	66	67
167.26	168.9342	173.04	174.97						
Er	Tm	Yb	Lu						
68	69	70	71						

Érbio
Erbium
Erbium
Erbio
эрбий

ארביום

鉺 エルビウム

Naturally occurring isotopes: 166, 168, 167, 170, 164, 162

Density: 9.066 g/cm^3 (25°C)

Melting point: 1529°C   **Boiling point:** 2863°C

Latent heat of fusion: 102.6 J/g

Specific heat: 0.168 J/g/°C (25°C)

Coefficient of lineal thermal expansion: 9.2 × 10^{-6} cm/cm/°C (25°C)

Thermal conductivity: 0.145 w/cm/°C (25°C)

Electrical resistivity: 107.0 × 10^{-6} ohm-cm (25°C)

Ionization potential (1st): 6.10 eV

Oxidation potential: $Er \rightarrow Er^{3+} + 3\epsilon = 2.296$ V

Chemical valence: 3

Electrochemical equivalents: 2.0802 g/amp-hr

Ionic radius: 0.881 Å (Er^{3+})

Valence electron potential ($-\epsilon V$): 49.0

Principal quantum number: 6

Principal electron shells: K L M N O P

Electronic configuration: $1s^2 2s^2 2p^6 3s^2 3p^6 3d^{10} 4s^2 4p^6 4d^{10} 4f^{12} 5s^2$
        $5p^6 6s^2$

Valence electrons: $4f^{12} 6s^2$

Crystal form: Hexagonal, close packed

Cross section σ: 160 ± 30 barns

Europium

63

151.96

Lanthanide Series

140.12 Ce 58	140.9077 Pr 59	144.24 Nd 60	144.913 Pm 61	150.4 Sm 62	151.96 Eu 63	157.25 Gd 64	158.9254 Tb 65	162.50 Dy 66	164.9304 Ho 67
167.26 Er 68	168.9342 Tm 69	173.04 Yb 70	174.97 Lu 71						

Európio
Europium
Europium
Europio
европий
אירופיום

銪 ユーロビウム

Naturally occurring isotopes: 153, 151
Density: 5.243 g/cm^3 (25°C)
Melting point: 822°C   **Boiling point:** 1597°C
Latent heat of fusion: 68.9 J/g
Specific heat: 0.182 J/g/°C (25°C)
Coefficient of lineal thermal expansion: 26 × 10^{-6} cm/cm/°C (20°C)
Thermal conductivity: 0.139 w/cm/°C (25°C)
Electrical resistivity: 81 × 10^{-6} ohm-cm (25°C)
Ionization potential (1st): 5.666 eV
Electron work function ϕ: 2.5 eV
Oxidation potential: Eu → Eu^{3+} + 3ϵ = 2.407 V
Chemical valence: 2, *3*
Electrochemical equivalents: 1.8899 g/amp-hr
Ionic radius: 0.947 Å (Eu^{3+})
Valence electron potential ($-\epsilon$V): 45.6
Principal quantum number: 6
Principal electron shells: K L M N O P
Electronic configuration: 1s^2 2s^2 2p^6 3s^2 3p^6 3d^{10} 4s^2 4p^6 4d^{10} 4f^7 5s^2
    5p^6 6s^2
Valence electrons: 4f^7 6s^2
Crystal form: Cubic, body centered
Cross section σ: 4100 ± 100 barns
Vapor pressure: 1.44 × 10^2 Pa (at melting point)

 **Fermium**

100

257.09515

Actinide Series

232.03807	231.0359	238.029	237.0482	244.06423	243.0614	247.07038	247.07032	251.07961	254.08805
Th	Pa	U	Np	Pu	Am	Cm	Bk	Cf	Es
90	91	92	93	94	95	96	97	98	99
257.09515	258	259	260						
Fm	Md	No	Lr						
100	101	102	103						

Férmio
Fermium
Fermium
Fermio
фермий
סרמיום
鐨 フェリミウム

Naturally occurring isotopes: None

Ionization potential (1st): 6.50 eV

Oxidation potential: $Fm \rightarrow Fm^{3+} + 3\epsilon = 2.0$ V

Chemical valence: 2, 3

Electrochemical equivalents: 3.1974 g/amp-hr

Principal quantum number: 7

Principal electron shells: K L M N O P Q

Electronic configuration: $1s^2 \ 2s^2 \ 2p^6 \ 3s^2 \ 3p^6 \ 3d^{10} \ 4s^2 \ 4p^6 \ 4d^{10} \ 4f^{14} \ 5s^2 \ 5p^6$ $5d^{10} \ 5f^{12} \ 6s^2 \ 6p^6 \ 7s^2$

Valence electrons: $5f^{12} \ 7s^2$

Half life: 80 days

Fluorine

9

18.998403

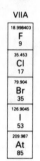

Flúor
Fluor
Fluor
Flúor
фтор
פלואור
氟 弗
素

Naturally occurring isotope: 19
Density: 1.696×10^{-3} g/cm^3 (0°C)
Melting point: −219.62°C   **Boiling point:** −188.14°C
Latent heat of fusion: 26.89 J/g (F$_2$)
Specific heat: 0.824 J/g/°C (F$_2$) (25°C)
Thermal conductivity: 0.279 mw/cm/°C (27°C at 1 atm)
Ionization potential (1st): 17.422 eV
Oxidation potential: $F^- \rightarrow \frac{1}{2}F_2 + \epsilon = -2.87$ V
Chemical valence: −1
Electrochemical equivalents: 0.70883 g/amp-hr
Ionic radius: 1.33 Å (F$^-$)
Valence electron potential ($-\epsilon V$): −10.1
Principal quantum number: 2
Principal electron shells: K L
Electronic configuration: 1s^2 2s^2 2p^5
Valence electrons: 2s^2 2p^5
Cross section σ: 9.8±0.7 mbarns
Vapor pressure: 4.90×10^2 Pa (at melting point)

Francium

87

223.01976

Francium

Frâncio
Francium
Franzium
Francio
франций
פרנציום
鈁 ワランシウム

IA
1.0079   .H   1
6.941   Li   3
22.98977   Na   11
39.098   K   19
85.4678   Rb   37
132.9054   Cs   55
223.01976   Fr   87

Naturally occurring isotopes: None (actinium decay product)
Melting point: 27°C (est)   **Boiling point:** 677°C (est)
Latent heat of fusion: 9.39 J/g (est)
Ionization potential (1st): 3.83 eV
Chemical valence: 1
Electrochemical equivalents: 8.3209 g/amp-hr
Ionic radius: 1.80 Å (Fr$^+$)
Valence electron potential ($-\varepsilon$V): 8.00
Principal quantum number: 7
Principal electron shells: K L M N O P Q
Electronic configuration: $1s^2\ 2s^2\ 2p^6\ 3s^2\ 3p^6\ 3d^{10}\ 4s^2\ 4p^6\ 4d^{10}\ 4f^{14}\ 5s^2\ 5p^6$
     $5d^{10}\ 6s^2\ 6p^6\ 7s^1$
Valence electrons: ($7s^1$)
Half life: 22 minutes
Crystal form: Cubic, body centered

Gadolinium

64

157.25

Lanthanide Series

Gadolínio
Gadolinium
Gadolinium
Gadolinio
гадолиний
גדוליניום
釓 ガドリニウム

140.12	140.9077	144.24	144.913	150.4	151.96	157.25	158.9254	162.50	164.9304
Ce	Pr	Nd	Pm	Sm	Eu	Gd	Tb	Dy	Ho
58	59	60	61	62	63	64	65	66	67
167.26	168.9342	173.04	174.97						
Er	Tm	Yb	Lu						
68	69	70	71						

Naturally occurring isotopes: 158, 160, 156, 157, 155, 154, 152
Density: 7.900 g/cm^3 (25°C)
Melting point: 1313°C   **Boiling point:** 3266°C
Latent heat of fusion: 98.51 J/g
Specific heat: 0.235 J/g/°C (25°C)
Coefficient of lineal thermal expansion: 9.7 $\times$ 10^{-6} cm/cm/°C (25°C)
Thermal conductivity: 0.105 w/cm/°C (25°C)
Electrical resistivity: 140.5 $\times$ 10^{-6} ohm-cm (25°C)
Ionization potential (1st): 6.14 eV
Electron work function ϕ: 3.1 eV
Oxidation potential: Gd $\rightarrow$ Gd^{3+} + 3ϵ = 2.397 V
Chemical valence: 3
Electrochemical equivalents: 1.9557 g/amp-hr
Ionic radius: 0.938 Å (Gd^{3+})
Valence electron potential ($-\epsilon$V): 46.1
Principal quantum number: 6
Principal electron shells: K L M N O P
Electronic configuration: 1s^2 2s^2 2p^6 3s^2 3p^6 3d^{10} 4s^2 4p^6 4d^{10} 4f^7 5s^2 5p^6
5d^1 6s^2
Valence electrons: 4f^7 5d^1 6s^2
Crystal form: Hexagonal, close packed
Cross section σ: 46,000 $\pm$ 2000 barns
Vapor pressure: 2.44 $\times$ 10^4 Pa (at melting point)

Ga Gallium

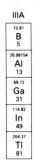

31

69.72

Gálio
Gallium
Gallium
Galio
галлий
גליום

鎵 ガリウム

Naturally occurring isotopes: 69, 71
Density: 5.906 g/cm^3 (25°C)
Melting point: 29.78°C   **Boiling point:** 2403°C
Latent heat of fusion: 80.17 J/g
Specific heat: 0.371 J/g/°C (25°C)
Coefficient of lineal thermal expansion: 18.1 × 10^{-6} cm/cm/°C (25°C)
Thermal conductivity: 0.281 w/cm/°C (liquid) (30°C)
Electrical resistivity: 17.4 × 10^{-6} ohm-cm (20°C)
Ionization potential (1st): 5.999 eV
Electron work function ϕ: 4.2 eV
Oxidation potential: Ga → Ga^{3+} + 3ϵ = −0.529 V
Chemical valence: 2, *3*
Electrochemical equivalents: 0.8671 g/amp-hr
Ionic radius: 0.620 Å (Ga^{3+})
Valence electron potential (−ϵV): 69.7
Principal quantum number: 4
Principal electron shells: K L M N
Electronic configuration: 1s^2 2s^2 2p^6 3s^2 3p^6 3d^{10} 4s^2 4p^1
Valence electrons: 4s^2 4p^1
Crystal form: Orthorhombic, rhombic
Cross section σ: 3.1 ± 0.3 barns
Vapor pressure: 9.31 × 10^{-36} Pa (at melting point)

Ge Germanium

32

72.59

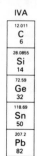

IVA

12.011 C 6
28.0855 Si 14
72.59 Ge 32
118.69 Sn 50
207.2 Pb 82

Germânio

Germanium

Germanium

Germanio

германий

גרמניום

鍺 ゲルマニウム

Naturally occurring isotopes: 74, 72, 70, 73, 76
Density: 5.323 g/cm³ (25°C)
Melting point: 937.4°C   **Boiling point:** 2830°C
Latent heat of fusion: 438.3 J/g
Specific heat: 0.3216 J/g/°C (25°C)
Coefficient of lineal thermal expansion: 5.75×10^{-6} cm/cm/°C (20°C)
Thermal conductivity: 0.667 w/cm/°C (25°C)
Electrical resistivity: 47 ohm-cm (intrinsic resistivity) (22°C)
Ionization potential (1st): 7.899 eV
Electron work function ϕ**:** 5.0 eV
Oxidation potential: $Ge + 2H_2O \rightarrow GeO_2 + 4H^+ + 4\epsilon = -0.15$ V
Chemical valence: $-4, 2, 4$
Electrochemical equivalents: 0.6771 g/amp-hr
Ionic radius: 0.530 Å (Ge^{4+})
Valence electron potential ($-\epsilon$**V):** 109
Principal quantum number: 4
Principal electron shells: K L M N
Electronic configuration: $1s^2\ 2s^2\ 2p^6\ 3s^2\ 3p^6\ 3d^{10}\ 4s^2\ 4p^2$
Valence electrons: $4s^2\ 4p^2$
Crystal form: Cubic, diamond
Cross section σ**:** 2.30 ± 0.26 barns
Vapor pressure: 7.46×10^{-5} Pa (at melting point)

Gold

79

196.9665

IB
63.546 **Cu** 29
107.868 **Ag** 47
196.9665 **Au** 79

Ouro

Or

Gold

Oro

золото

זהב

金 金

Naturally occurring isotope: 197
Density: 19.32 g/cm^3 (20°C)
Melting point: 1064.43°C   **Boiling point:** 3080°C
Latent heat of fusion: 62.81 J/g
Specific heat: 0.1290 J/g/°C (25°C)
Coefficient of lineal thermal expansion: 14.2 × 10^{-6} cm/cm/°C (20°C)
Thermal conductivity: 3.19 w/cm/°C (25°C)
Electrical resistivity: 2.44 × 10^{-6} ohm-cm (20°C)
Ionization potential (1st): 9.225 eV
Electron work function ϕ: 5.1 eV
Oxidation potentials: $Au \rightarrow Au^+ + \epsilon = -1.691$ V
$\qquad\qquad\qquad\quad\ Au \rightarrow Au^{3+} + 3\epsilon = -1.498$ V
Chemical valence: 1, *3*
Electrochemical equivalents: 2.4496 g/amp-hr
Ionic radius: 0.85 Å (Au^{3+})
Valence electron potential ($-\epsilon V$): 51
Principal quantum number: 6
Principal electron shells: K L M N O P
Electronic configuration: $1s^2\ 2s^2\ 2p^6\ 3s^2\ 3p^6\ 3d^{10}\ 4s^2\ 4p^6\ 4d^{10}\ 4f^{14}\ 5s^2\ 5p^6$
$\qquad\qquad 5d^{10}\ 6s^1$
Valence electrons: $5d^{10}\ 6s^1$
Crystal form: Cubic, face centered
Cross section σ: 98.8 ± 0.3 barns
Vapor pressure: 2.37 × 10^{-4} Pa (at melting point)

Hf Hafnium

72

178.49

IVB
47.90   Ti   22
91.22   Zr   40
178.49   Hf   72
104

Háfnio
Hafnium
Hafnium
Hafnio
гафний
המניום

鉿 ハフニウム

Naturally occurring isotopes: 180, 178, 177, 179, 176, 174

Density: 13.31 g/cm^3 (20°C)

Melting point: 2227 ± 20°C   **Boiling point:** 4602°C

Latent heat of fusion: 122.0 J/g

Specific heat: 0.144 J/g/°C (25°C)

Coefficient of lineal thermal expansion: 5.6 × 10^{-6} cm/cm/°C (25°C)

Thermal conductivity: 0.230 w/cm/°C (25°C)

Electrical resistivity: 35.1 × 10^{-6} ohm-cm (25°C)

Ionization potential (1st): 6.65 eV

Electron work function ϕ: 3.9 eV

Oxidation potential: Hf → Hf^{4+} + 4ϵ = 1.70 V

Chemical valence: 4

Electrochemical equivalents: 1.6649 g/amp-hr

Ionic radius: 0.71 Å (Hf^{4+})

Valence electron potential ($-\epsilon$V): 81

Principal quantum number: 6

Principal electron shells: K L M N O P

Electronic configuration: 1s^2 2s^2 2p^6 3s^2 3p^6 3d^{10} 4s^2 4p^6 4d^{10} 4f^{14} 5s^2 5p^6 5d^2 6s^2

Valence electrons: 5d^2 6s^2

Crystal form: Hexagonal, close packed

Cross section σ: 103 ± 3 barns

Vapor pressure: 1.12 × 10^{-3} Pa (at melting point)

Helium

2

4.00260

	O	
	4.00260	
	He	
	2	
	20.179	
	Ne	
	10	
	39.948	
	Ar	
	18	
	83.80	
	Kr	
	36	
	131.30	
	Xe	
	54	
	222.01761	
	Rn	
	86	

Hélio

Hélium

Helium

Helio

гелий

הליום

氦 ヘリウム

Naturally occurring isotopes: 4, 3
Density: 0.17847×10^{-3} g/cm^3 (0°C)
Melting point: -272.2°C (26 atm); I **Boiling point:** -268.934°C
Latent heat of fusion: 5.23 J/g
Specific heat: 5.1930 J/g/°C (25°C)
Thermal conductivity: 1.520 mw/cm/°C (25°C at 1 atm)
Ionization potential (1st): 24.58 eV
Chemical valence: 0
Principal quantum number: 1
Principal electron shells: K
Electronic configuration: 1s^2
Valence electrons: (1s^2)
Crystal form: Hexagonal, close packed
Cross section σ: 0.007 barns

Ho Holmium

67

164.9304

Lanthanide Series

140.12	140.9077	144.24	144.913	150.4	151.96	157.25	158.9254	162.50	164.9304
Ce	Pr	Nd	Pm	Sm	Eu	Gd	Tb	Dy	Ho
58	59	60	61	62	63	64	65	66	67
167.26	168.9342	173.04	174.97						
Er	Tm	Yb	Lu						
68	69	70	71						

Naturally occurring isotope: 165

Density: 8.795 g/cm³ (25°C)

Melting point: 1474°C   **Boiling point:** 2695°C

Latent heat of fusion: 104.1 J/g

Specific heat: 0.165 J/g/°C (25°C)

Coefficient of lineal thermal expansion: 9.5×10^{-6} cm/cm/°C (400°C)

Thermal conductivity: 0.162 w/cm/°C (25°C)

Electrical resistivity: 87.0×10^{-6} ohm-cm (25°C)

Ionization potential (1st): 6.02 eV

Oxidation potential: $Ho \rightarrow Ho^{3+} + 3\epsilon = 2.319$ V

Chemical valence: 3

Electrochemical equivalents: 2.0512 g/amp-hr

Ionic radius: 0.901 Å (Ho^{3+})

Valence electron potential ($-\epsilon V$): 47.9

Principal quantum number: 6

Principal electron shells: K L M N O P

Electronic configuration: $1s^2\ 2s^2\ 2p^6\ 3s^2\ 3p^6\ 3d^{10}\ 4s^2\ 4p^6\ 4d^{10}\ 4f^{11}\ 5s^2\ 5p^6\ 6s^2$

Valence electrons: $4f^{11}\ 6s^2$

Crystal form: Hexagonal, close packed

Cross section σ: 65 ± 2 barns

Hydrogen

1

1.0079

IA

1.0079	
H	
1	

| 6.941 |
| Li |
| 3 |

| 22.98977 |
| Na |
| 11 |

| 39.098 |
| K |
| 19 |

| 85.4678 |
| Rb |
| 37 |

| 132.9054 |
| Cs |
| 55 |

| 223.01976 |
| Fr |
| 87 |

Hidrogênio
Hydrogène
Wasserstoff
Hidrógeno
водород
מימן
氫 水素

Naturally occurring isotopes: 1.007825 (protium), 2.01410 (deuterium), 3.01605 (tritium)

Density: 0.08988×10^{-3} g/cm^3 (0°C)

Melting point: -259.14°C   **Boiling point:** -252.87°C

Latent heat of fusion: 116.3 J/g (H$_2$)

Specific heat: 14.30 J/g/°C (H$_2$) (25°C)

Thermal conductivity: 1.815 mw/cm/°C (27°C at 1 atm)

Ionization potential (1st): 13.59765 eV

Oxidation potentials: $H_2 \rightarrow 2H^+ + \epsilon = 0.00000$ V
$$H^- \rightarrow \tfrac{1}{2}H_2 + \epsilon = 2.25 \text{ V}$$

Chemical valence: 1

Electrochemical equivalents: 0.037605 g/amp-hr

Ionic radius: 0.012 Å (H$^+$)

Valence electron potential ($-\epsilon$V): 1200

Principal quantum number: 1

Principal electron shells: K

Electronic configuration: 1s^1

Valence electrons: 1s^1

Crystal form: Hexagonal, close packed

Cross section σ: 0.33 barns

Indium

49

114.82

IIIA
10.81 **B** 5
26.98154 **Al** 13
69.72 **Ga** 31
114.82 **In** 49
204.37 **Tl** 81

Índio
Indium
Indium
Indio
индий
אינדיום

鈿 イソジウム

Naturally occurring isotopes: 115, 113
Density: 7.28 g/cm^3 (20°C)
Melting point: 156.61°C **Boiling point:** 2080°C
Latent heat of fusion: 28.44 J/g
Specific heat: 0.233 J/g/°C (25°C)
Coefficient of lineal thermal expansion: 24.8 × 10^{-6} cm/cm/°C (20°C)
Thermal conductivity: 0.818 w/cm/°C (25°C)
Electrical resistivity: 8.37 × 10^{-6} ohm-cm (0°C)
Ionization potential (1st): 5.786 eV
Electron work function ϕ: 4.12 eV
Oxidation potential: In → In^{3+} + 3ϵ = 0.343 V
Chemical valence: 1, 2, *3*
Electrochemical equivalents: 1.4280 g/amp-hr
Ionic radius: 0.800 Å (In^{3+})
Valence electron potential ($-\epsilon$V): 54.0
Principal quantum number: 5
Principal electron shells: K L M N O
Electronic configuration: 1s^2 2s^2 2p^6 3s^2 3p^6 3d^{10} 4s^2 4p^6 4d^{10} 5s^2 5p^1
Valence electrons: 5s^2 5p^1
Crystal form: Tetragonal
Cross section σ: 194 ± 2 barns
Vapor pressure: 1.42 × 10^{17} Pa (at melting point)

Iodine

VIIA		
18.998403	F	9
35.453	Cl	17
79.904	Br	35
126.9045	I	53
209.987	At	85

53

126.9045

Iôdo

Iode

Iod

Yodo

иод

יוד

碘 沃
素

Naturally occurring isotope: 127

Density: 4.93 g/cm^3 (20°C)

Melting point: 113.5°C   **Boiling point:** 184.35°C

Latent heat of fusion: 124.4 J/g (I$_2$)

Specific heat: 0.21448 J/g/°C (25°C)

Coefficient of lineal thermal expansion: 93 $\times$ 10^{-6} cm/cm/°C (20°C)

Thermal conductivity: 4.49 mw/cm/°C (27°C)

Electrical resistivity: 1.3 $\times$ 10^9 ohm-cm (20°C)

Ionization potential (1st): 10.451 eV

Oxidation potential: I$^-$ $\rightarrow$ ½I$_2$ + ϵ = $-$0.5355 V

Chemical valence: $-$ 1, 3, 5, 7

Electrochemical equivalents: 4.7348 g/amp-hr

Ionic radius: 2.20 Å (I$^-$)

Valence electron potential ($-\epsilon$V): $-$6.55

Principal quantum number: 5

Principal electron shells: K L M N O

Electronic configuration: 1s^2 2s^2 2p^6 3s^2 3p^6 3d^{10} 4s^2 4p^6 4d^{10} 5s^2 5p^5

Valence electrons: 5s^2 5p^5

Crystal form: Orthorhombic

Cross section σ: 6.2 $\pm$ 0.2 barns

Ir Iridium

77

192.22

VIII		
55.847 Fe 26	58.9332 Co 27	58.70 Ni 28
101.07 Ru 44	102.9055 Rh 45	106.4 Pd 46
190.2 Os 76	192.22 Ir 77	195.09 Pt 78
	109	

Irídio

Iridium

Iridium

Iridio

иридий

אירידיום

鉄 イリジウム

Naturally occurring isotopes: 193, 191
Density: 22.42 g/cm^3 (17°C)
Melting point: 2410°C   **Boiling point:** 4130°C
Latent heat of fusion: 137.2 J/g
Specific heat: 0.131 J/g/°C (25°C)
Coefficient of lineal thermal expansion: 6.6 × 10^{-6} cm/cm/°C (25°C)
Thermal conductivity: 1.47 w/cm/°C (25°C)
Electrical resistivity: 4.71 × 10^{-6} ohm-cm (20°C)
Ionization potential (1st): 9.1 eV
Electron work function φ: 5.27 eV
Oxidation potential: $Ir + 6Cl^- \rightarrow IrCl_6^{3-} + 3\epsilon = -0.77$ V
Chemical valence: 2, 3, *4*, 6
Electrochemical equivalents: 1.793 g/amp-hr
Ionic radius: 0.625 Å (Ir^{4+})
Valence electron potential ($-\epsilon$V): 92.2
Principal quantum number: 6
Principal electron shells: K L M N O P
Electronic configuration: $1s^2\ 2s^2\ 2p^6\ 3s^2\ 3p^6\ 3d^{10}\ 4s^2\ 4p^6\ 4d^{10}\ 4f^{14}\ 5s^2\ 5p^6$
    $5d^7\ 6s^2$
Valence electrons: $5d^7\ 6s^2$
Crystal form: Cubic, face centered
Cross section σ: 425 ± 15 barns
Vapor pressure: 1.47 Pa (at melting point)

Fe Iron

26

55.847

	VIII	
55.847 Fe 26	58.9332 Co 27	58.70 Ni 28
101.07 Ru 44	102.9055 Rh 45	106.4 Pd 46
190.2 Os 76	192.22 Ir 77	195.09 Pt 78
	109	

Ferro

Fer

Eisen

Hierro

железо

ברזל

鉄 鉄

Naturally occurring isotopes: 56, 54, 57, 58
Density: 7.874 g/cm^3 (20°C)
Melting point: 1535°C   **Boiling point:** 2750°C
Latent heat of fusion: 275.1 J/g
Specific heat: 0.450 J/g/°C (25°C)
Coefficient of lineal thermal expansion: 11.76 × 10^{-6} cm/cm/°C (20°C)
Thermal conductivity: 0.804 w/cm/°C (25°C)
Electrical resistivity: 9.71 × 10^{-6} ohm-cm (20°C)
Ionization potential (1st): 7.870 eV
Electron work function ϕ: 4.70 eV
Oxidation potential: $Fe \rightarrow Fe^{2+} + 2\epsilon = 0.4402$ V
Chemical valence: 2, *3*, 4, 6
Electrochemical equivalents: 0.69455 g/amp-hr
Ionic radius: 0.645 Å (Fe^{3+})
Valence electron potential (−εV): 67.0
Principal quantum number: 4
Principal electron shells: K L M N
Electronic configuration: $1s^2\ 2s^2\ 2p^6\ 3s^2\ 3p^6\ 3d^6\ 4s^2$
Valence electrons: $3d^6\ 4s^2$
Crystal form: Cubic, body centered
Cross section σ: 2.56 ± 0.05 barns
Vapor pressure: 7.05 Pa (at melting point)

Kr Krypton

36

83.80

O	
4.00260 He 2	Criptônio
20.179 Ne 10	Krypton
39.948 Ar 18	Krypton
83.80 Kr 36	Criptón
131.30 Xe 54	криптон
222.01761 Rn 86	קריסטון

氪 ワリプトン

Naturally occurring isotopes: 84, 86, 82, 83, 80, 78
Density: 3.733×10^{-3} g/cm³ (20°C)
Melting point: -156.6°C   **Boiling point:** -152.30 ± 0.10°C
Latent heat of fusion: 19.54 J/g
Specific heat: 0.24804 J/g/°C (25°C)
Thermal conductivity: 0.0949 mw/cm/°C (27°C)
Ionization potential (1st): 13.999 eV
Chemical valence: 0
Principal quantum number: 4
Principal electron shells: K L M N
Electronic configuration: $1s^2\ 2s^2\ 2p^6\ 3s^2\ 3p^6\ 3d^{10}\ 4s^2\ 4p^6$
Valence electrons: $(4s^2\ 4p^6)$
Crystal form: Cubic, face centered (solid)
Cross section σ: 24.5 ± 1.0 barns

Lanthanum

57

138.9055

IIIB

| 44.95592 |
| Sc |
| 21 |
| 88.9059 |
| Y |
| 39 |
| 138.9055 |
| La |
| 57 |
| 227.02777 |
| Ac |
| 89 |

Lantânio
Lanthane
Lanthan
Lantano
лантан

לנתן

Naturally occurring isotopes: 139, 138
Density: 6.145 g/cm^3 (25°C)
Melting point: 921°C   **Boiling point:** 3457°C
Latent heat of fusion: 81.4 J/g
Specific heat: 0.195 J/g/°C (25°C)
Coefficient of lineal thermal expansion: 5.2 × 10^{-6} cm/cm/°C (25°C)
Thermal conductivity: 0.134 w/cm/°C (25°C)
Electrical resistivity: 56 × 10^{-6} ohm-cm (25°C)
Ionization potential (1st): 5.577 eV
Electron work function ϕ: 3.5 eV
Oxidation potential: La → La^{3+} + 3ϵ = 2.522 V
Chemical valence: 3
Electrochemical equivalents: 1.7275 g/amp-hr
Ionic radius: 1.061 Å (La^{3+})
Valence electron potential ($-\epsilon$V): 40.71
Principal quantum number: 6
Principal electron shells: K L M N O P
Electronic configuration: 1s^2 2s^2 2p^6 3s^2 3p^6 3d^{10} 4s^2 4p^6 4d^{10} 5s^2 5p^6
    5d^1 6s^2
Valence electrons: 5d^1 6s^2
Crystal form: Hexagonal, close packed
Cross section σ: 8.9 ± 0.2 barns
Vapor pressure: 1.33 × 10^{-7} Pa (at melting point)

Lawrencium

103

260

Actinide Series

Laurêncio
Lawrencium
Lawrenzium
Lawrencio
лавренций

テーレンチウム

232.03807	231.0359	238.029	237.0482	244.06423	243.0614	247.07038	247.07032	251.07961	254.08805
Th	Pa	U	Np	Pu	Am	Cm	Bk	Cf	Es
90	91	92	93	94	95	96	97	98	99
257.09515	258	259	260						
Fm	Md	No	Lr						
100	101	102	103						

Naturally occurring isotopes: None

Oxidation potential: $Lr \rightarrow Lr^{3+} + 3\epsilon = 2.0$ V

Chemical valence: 3

Electrochemical equivalents: 3.23 g/amp-hr

Principal quantum number: 7

Principal electron shells: K L M N O P Q

Electronic configuration: $1s^2\ 2s^2\ 2p^6\ 3s^2\ 3p^6\ 3d^{10}\ 4s^2\ 4p^6\ 4d^{10}\ 4f^{14}\ 5s^2\ 5p^6$ $5d^{10}\ 5f^{14}\ 6s^2\ 6p^6\ 6d^1\ 7s^2$

Valence electrons: $5f^{14}\ 6d^1\ 7s^2$

Half life: 3 minutes

Lead

82

207.2

IVA

| 12.011 C 6 |
| 28.0855 Si 14 |
| 72.59 Ge 32 |
| 118.69 Sn 50 |
| 207.2 Pb 82 |

Chumbo

Plomb

Blei

Plomo

свинец

עופרת

鉛 鉛

Naturally occurring isotopes: 208, 206, 207, 204
Density: 11.342 g/cm^3 (20°C)
Melting point: 327.502°C   **Boiling point:** 1740°C
Latent heat of fusion: 23.06 J/g
Specific heat: 0.128 J/g/°C (25°C)
Coefficient of lineal thermal expansion: 28.3 × 10^{-6} cm/cm/°C (25°C)
Thermal conductivity: 0.353 w/cm/°C (25°C)
Electrical resistivity: 20.65 × 10^{-6} ohm-cm (20°C)
Ionization potential (1st): 7.416 eV
Electron work function φ: 4.25 eV
Oxidation potential: Pb → Pb^{2+} + 2ε = 0.126 V
Chemical valence: 2, 4
Electrochemical equivalents: 3.865 g/amp-hr (Pb^{2+})
Ionic radius: 1.19 Å (Pb^{2+})
Valence electron potential (−εV): 24.2
Principal quantum number: 6
Principal electron shells: K L M N O P
Electronic configuration: 1s^2 2s^2 2p^6 3s^2 3p^6 3d^{10} 4s^2 4p^6 4d^{10} 4f^{14} 5s^2 5p^6 5d^{10} 6s^2 6p^2
Valence electrons: 6s^2 6p^2
Crystal form: Cubic, face centered
Cross section σ: 180 ± 10 mbarns
Vapor pressure: 4.21 × 10^{-7} Pa (at melting point)

---

47   LEAD [PLUMBUM]

Lithium

3

6.941

IA

1.0079	H 1
6.941	Li 3
22.98977	Na 11
39.098	K 19
85.4678	Rb 37
132.9054	Cs 55
223.01976	Fr 87

Litio
Lithium
Lithium
Litio
литий
ליתיום

鋰 リチウム

Naturally occurring isotopes: 7, 6
Density: 0.534 g/cm^3 (20°C)
Melting point: 180.54°C   **Boiling point:** 1342°C
Latent heat of fusion: 430.1 J/g
Specific heat: 3.57 J/g/°C (25°C)
Coefficient of lineal thermal expansion: 60 × 10^{-6} cm/cm/°C (25°C)
Thermal conductivity: 0.848 w/cm/°C (25°C)
Electrical resistivity: 8.55 × 10^{-6} ohm-cm (0°C)
Ionization potential (1st): 5.392 eV
Electron work function ϕ: 2.9 eV
Oxidation potential: Li → Li$^+$ + ϵ = 3.045 V
Chemical valence: 1
Electrochemical equivalents: 0.2590 g/amp-hr
Ionic radius: 0.76 Å (Li$^+$)
Valence electron potential ($-\epsilon$V): 19
Principal quantum number: 2
Principal electron shells: K L
Electronic configuration: 1s^2 2s^1
Valence electrons: 2s^1
Crystal form: Cubic, body centered
Cross section σ: 71 barns
Vapor pressure: 1.63 × 10^{-8} Pa (at melting point)

Lu **Lutetium**

71

174.97

Lanthanide Series

Lutécio
Lutetium
Lutetium
Lutecio
лютеций
לוטציום

鑥 ルテチウム

140.12	140.9077	144.24	144.913	150.4	151.96	157.25	158.9254	162.50	164.9304
Ce	Pr	Nd	Pm	Sm	Eu	Gd	Tb	Dy	Ho
58	59	60	61	62	63	64	65	66	67
167.26	168.9342	173.04	174.97						
Er	Tm	Yb	Lu						
68	69	70	71						

Naturally occurring isotopes: 175, 176
Density: 9.840 g/cm^3 (25°C)
Melting point: 1663°C   **Boiling point:** 3395°C
Latent heat of fusion: 110.1 J/g
Specific heat: 0.154 J/g/°C (25°C)
Coefficient of lineal thermal expansion: 12.5 × 10^{-6} cm/cm/°C (400°C)
Thermal conductivity: 0.164 w/cm/°C (25°C)
Electrical resistivity: 79.0 × 10^{-6} ohm-cm (25°C)
Ionization potential (1st): 5.4259 eV
Electron work function ϕ: 3.3 eV
Oxidation potential: Lu → Lu^{3+} + 3ϵ = 2.255 V
Chemical valence: 3
Electrochemical equivalents: 2.1760 g/amp-hr
Ionic radius: 0.848 Å (Lu^{3+})
Valence electron potential ($-\epsilon$V): 50.9
Principal quantum number: 6
Principal electron shells: K L M N O P
Electronic configuration: 1s^2 2s^2 2p^6 3s^2 3p^6 3d^{10} 4s^2 4p^6 4d^{10} 4f^{14} 5s^2 5p^6
     5d^1 6s^2
Valence electrons: 5d^1 6s^2
Crystal form: Hexagonal, close packed
Cross section σ: 75±2 barns
Vapor pressure: 2.46 × 10^3 Pa (at melting point)

 **Magnesium**

	IIA
12	9.01218 **Be** 4
	24.305 **Mg** 12
24.305	40.08 **Ca** 20
	87.62 **Sr** 38
	137.34 **Ba** 56
	226.02544 **Ra** 88

Magnésio
Magnésium
Magnesium
Magnesio
магний
מגנזיום
鎂 マグネシウム

Naturally occurring isotopes: 24, 26, 25
Density: 1.738 g/cm^3 (20°C)
Melting point: 648.8 ± 0.5°C   **Boiling point:** 1090°C
Latent heat of fusion: 368.6 J/g
Specific heat: 102 J/g/°C (25°C)
Coefficient of lineal thermal expansion: 27.1 × 10^{-6} cm/cm/°C (20°C)
Thermal conductivity: 1.56 w/cm/°C (20°C)
Electrical resistivity: 4.45 × 10^{-6} ohm-cm (20°C)
Ionization potential (1st): 7.646 eV
Electron work function ϕ: 3.66 eV
Oxidation potential: Mg → Mg^{2+} + 2ϵ = 2.363 V
Chemical valence: 2
Electrochemical equivalents: 0.45341 g/amp-hr
Ionic radius: 0.72 Å (Mg^{2+})
Valence electron potential ($-\epsilon$V): 40
Principal quantum number: 3
Principal electron shells: K L M
Electronic configuration: 1s^2 2s^2 2p^6 3s^2
Valence electrons: 3s^2
Crystal form: Hexagonal, close packed
Cross section σ: 64 ± 2 mbarns
Vapor pressure: 3.61 × 10^2 (at melting point)

 **Manganese**

25

54.9380

	VIIB	
	54.9380 **Mn** 25	
	96.906 **Tc** 43	
	186.2 **Re** 75	
	107	

Manganês

Manganese

Mangan

Manganeso

марганец

מנגן

鑑 ヌンガン

Naturally occurring isotope: 55
Density: 7.44 g/cm^3 (20°C)
Melting point: 1244±3°C **Boiling point:** 1962°C
Latent heat of fusion: 266.7 J/g
Specific heat: 0.479 J/g/°C (20°C)
Coefficient of lineal thermal expansion: 22 × 10^{-6} cm/cm/°C (20°C)
Thermal conductivity: 78.1 mw/cm/°C (25°C)
Electrical resistivity: 185 × 10^{-6} ohm-cm (20°C)
Ionization potential (1st): 7.435 eV
Electron work function ϕ: 4.1 eV
Oxidation potential: Mn → Mn^{2+} + 2ϵ = 1.18 V
Chemical valence: −2, −1, 0, 1, 2, 3, 4, 5, 6, 7
Electrochemical equivalents: 0.29282 g/amp-hr
Ionic radius: 0.46 Å (Mn^{7+})
Valence electron potential (−ϵV): 220
Principal quantum number: 4
Principal electron shells: K L M N
Electronic configuration: 1s^2 2s^2 2p^6 3s^2 3p^6 3d^5 4s^2
Valence electrons: 3d^5 4s^2
Crystal form: Cubic, face centered
Cross section σ: 13.3±0.1 barns
Vapor pressure: 1.21 × 10^2 Pa (at melting point)

 **Mendelevium**

101

258

Actinide Series

232.03807	231.0359	238.029	237.0482	244.06423	243.0614	247.07038	247.07032	251.07961	254.08805
Th 90	Pa 91	U 92	Np 93	Pu 94	Am 95	Cm 96	Bk 97	Cf 98	Es 99

257.09515	258	259	260
Fm 100	Md 101	No 102	Lr 103

Naturally occurring isotopes: None

Ionization potential (1st): 6.58 eV

Oxidation potential: $Md \rightarrow Md^{3+} + 3\epsilon = 1.6$ V

Chemical valence: 1, 2, *3*

Electrochemical equivalents: 3.21 g/amp-hr

Principal quantum number: 7

Principal electron shells: K L M N O P Q

Electronic configuration: $1s^2\ 2s^2\ 2p^6\ 3s^2\ 3p^6\ 3d^{10}\ 4s^2\ 4p^6\ 4d^{10}\ 4f^{14}\ 5s^2\ 5p^6$ $5d^{10}\ 5f^{13}\ 6s^2\ 6p^6\ 7s^2$

Valence electrons: $5f^{13}\ 7s^2$

Half life: 55 days

Mercury

80

200.59

IIB

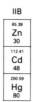

| 65.38 |
| Zn |
| 30 |

| 112.41 |
| Cd |
| 48 |

| 200.59 |
| Hg |
| 80 |

Mercúrio

Mercure

Quecksilber

Mercurio

ртуть

כספית

汞 水銀

Naturally occurring isotopes: 202, 200, 199, 201, 198, 204, 196

Density: 13.534 g/cm³ (25°C)

Melting point: $-38.87°C$ **Boiling point:** 356.58°C

Latent heat of fusion: 11.46 J/g

Specific heat: 0.1395 J/g/°C (liquid) (25°C)

Thermal conductivity: 0.0830 w/cm/°C (25°C)

Electrical resistivity: 95.78×10^{-6} ohm-cm (20°C)

Ionization potential (1st): 10.437 eV

Electron work function ϕ: 4.49 eV

Oxidation potential: $Hg \rightarrow Hg^{2+} + 2\epsilon = -0.788$ V

Chemical valence: 1, *2*

Electrochemical equivalents: 3.7420 g/amp-hr

Ionic radius: 1.02 Å (Hg^{2+})

Valence electron potential $(-\epsilon V)$: 28.2

Principal quantum number: 6

Principal electron shells: K L M N O P

Electronic configuration: $1s^2\ 2s^2\ 2p^6\ 3s^2\ 3p^6\ 3d^{10}\ 4s^2\ 4p^6\ 4d^{10}\ 4f^{14}\ 5s^2\ 5p^6$
    $5d^{10}\ 6s^2$

Valence electrons: $6s^2$

Crystal form: Rhombohedral

Cross section σ: 375 ± 5 barns

Vapor pressure: 2.00×10^{-4} Pa (at melting point)

Mo  Molybdenum

42
95.94

VIB
51.996 Cr 24
95.94 Mo 42
183.85 W 74
106

Molibdênio

Molybdène

Molybdän

Molibdeno

молибден

מוליבדן

Naturally occurring isotopes: 98, 96, 95, 92, 100, 97, 94
Density: 10.22 g/cm^3 (20°C)
Melting point: 2617°C   Boiling point: 4612°C
Latent heat of fusion: 288.0 J/g
Specific heat: 0.251 J/g/°C (25°C)
Coefficient of lineal thermal expansion: 6.6 × 10^{-6} cm/cm/°C (25°C)
Thermal conductivity: 1.38 w/cm/°C (25°C)
Electrical resistivity: 5.2 × 10^{-6} ohm-cm (0°C)
Ionization potential (1st): 7.099 eV
Electron work function φ: 4.6 eV
Oxidation potential: $Mo \rightarrow Mo^{3+} + 3\epsilon = 0.2$ V
Chemical valence: 2, 3, *4*, 5, 6
Electrochemical equivalents: 0.8949 g/amp-hr
Ionic radius: 0.650 Å (Mo^{4+})
Valence electron potential (−εV): 88.6
Principal quantum number: 5
Principal electron shells: K L M N O
Electronic configuration: $1s^2 2s^2 2p^6 3s^2 3p^6 3d^{10} 4s^2 4p^6 4d^5 5s^1$
Valence electrons: $4d^5 5s^1$
Crystal form: Cubic, body centered
Cross section σ: 2.65 ± 0.05 barns
Vapor pressure: 3.47 Pa (at melting point)

Neodymium

60

144.24

Lanthanide Series

Neodímio
Neodymium
Neodym
Neodimio
неодимий
ניאודימיום

140.12	140.9077	144.24	144.913	150.4	151.96	157.25	158.9254	162.50	164.9304
Ce	Pr	Nd	Pm	Sm	Eu	Gd	Tb	Dy	Ho
58	59	60	61	62	63	64	65	66	67
167.26	168.9342	173.04	174.97						
Er	Tm	Yb	Lu						
68	69	70	71						

Naturally occurring isotopes: 142, 144, 146, 143, 145, 148, 150

Density: 7.007 g/cm^3 (25°C)

Melting point: 1021°C   **Boiling point:** 3068°C

Latent heat of fusion: 75.47 J/g

Specific heat: 0.190 J/g/°C (25°C)

Coefficient of lineal thermal expansion: 8.6 × 10^6 cm/cm/°C (25°C)

Thermal conductivity: 0.165 w/cm/°C (25°C)

Electrical resistivity: 64.0 × 10^{-6} ohm-cm (25°C)

Ionization potential (1st): 5.49 eV

Electron work function ϕ: 3.2 eV

Oxidation potential: Nd → Nd^{3+} + 3ϵ = 2.431 V

Chemical valence: 2, *3*

Electrochemical equivalents: 1.7939 g/amp-hr

Ionic radius: 0.995 Å (Nd^{3+})

Valence electron potential ($-\epsilon$V): 43.4

Principal quantum number: 6

Principal electron shells: K L M N O P

Electronic configuration: 1s^2 2s^2 2p^6 3s^2 3p^6 3d^{10} 4s^2 4p^6 4d^{10} 4f^4 5s^2 5p^6 6s^2

Valence electrons: 4f^4 6s^2

Crystal form: Hexagonal, close packed

Cross section σ: 49±2 barns

Vapor pressure: 6.03 × 10^{-3} Pa (at melting point)

Neon

10

20.179

	O	
	4.00260	
	He	
	2	
	20.179	
	Ne	
	10	
	39.948	
	Ar	
	18	
	83.80	
	Kr	
	36	
	131.30	
	Xe	
	54	
	222.01761	
	Rn	
	86	

Neônio

Neon

Neon

Neón

неон

ניאון

気 ネオン

Naturally occurring isotopes: 20, 22, 21
Density: 0.8999×10^{-3} g/cm^3 (20°C)
Melting point: -248.67°C   **Boiling point:** -246.048°C
Latent heat of fusion: 16.6 J/g
Specific heat: 1.0301 J/g/°C (25°C)
Thermal conductivity: 0.493 mw/cm/°C (27°C)
Ionization potential (1st): 21.564 eV
Chemical valence: 0
Principal quantum number: 2
Principal electron shells: K L
Electronic configuration: $1s^2\ 2s^2\ 2p^6$
Valence electrons: ($2s^2\ 2p^6$)
Crystal form: Cubic, face centered
Cross section σ: 38 ± 10 mbarns

Neptunium

93

237.0482

Actinide Series

Neptúnio
Neptunium
Neptunium
Neptunio
нептуний
נפטוניום
鎿 ネプツニウム

232.03807	231.0359	238.029	237.0482	244.06423	243.0614	247.07038	247.07032	251.07961	254.08805
Th	Pa	U	Np	Pu	Am	Cm	Bk	Cf	Es
90	91	92	93	94	95	96	97	98	99
257.09515	258	259	260						
Fm	Md	No	Lr						
100	101	102	103						

Naturally occurring isotopes: None
Density: 20.45 g/cm^3 (25°C)
Melting point: 640 ± 1°C   **Boiling point:** 3902°C
Latent heat of fusion: 46 J/g
Thermal conductivity: 63 mw/cm/°C (27°C)
Electrical resistivity: 119 × 10^{-6} ohm-cm (100°C)
Ionization potential (1st): 6.19 eV
Oxidation potential: $Np \rightarrow Np^{3+} + 3\epsilon = 1.856$ V
Chemical valence: 3, 4, *5,* 6, 7
Electrochemical equivalents: 1.7689 g/amp-hr
Ionic radius: 0.75 Å (Np^{5+})
Valence electron potential ($-\epsilon V$): 96
Principal quantum number: 7
Principal electron shells: K L M N O P Q
Electronic configuration: $1s^2\ 2s^2\ 2p^6\ 3s^2\ 3p^6\ 3d^{10}\ 4s^2\ 4p^6\ 4d^{10}\ 4f^{14}\ 5s^2\ 5p^6$
    $5d^{10}\ 5f^4\ 6s^2\ 6p^6\ 6d^1\ 7s^2$
Valence electrons: $5f^4\ 6d^1\ 7s^2$
Crystal form: Orthorhombic
Half life: 2.14 × 10^4 years
Cross section σ: 170 ± 5 barns

Nickel

28

58.70

	VIII	
55.847 **Fe** 26	58.9332 **Co** 27	58.70 **Ni** 28
101.07 **Ru** 44	102.9055 **Rh** 45	106.4 **Pd** 46
190.2 **Os** 76	192.22 **Ir** 77	195.09 **Pt** 78
	109	

Niquel

Nickel

Nickel

Niquel

никель

ניקל

鎳 ニッケル

Naturally occurring isotopes: 58, 60, 62, 61, 64
Density: 8.902 g/cm^3 (25°C)
Melting point: 1453°C   **Boiling point:** 2732°C
Latent heat of fusion: 300.3 J/g
Specific heat: 0.444 J/g/°C (25°C)
Coefficient of lineal thermal expansion: 13.3 × 10^{-6} cm/cm/°C (20°C)
Thermal conductivity: 0.909 w/cm/°C (25°C)
Electrical resistivity: 6.84 × 10^{-6} ohm-cm (20°C)
Ionization potential (1st): 7.635 eV
Electron work function ϕ: 5.15 eV
Oxidation potential: Ni → Ni^{2+} + 2ϵ = 0.250 V
Chemical valence: 0, 1, *2*, 3
Electrochemical equivalents: 1.095 g/amp-hr
Ionic radius: 0.69 Å (Ni^{2+})
Valence electron potential ($-\epsilon$V): 42
Principal quantum number: 4
Principal electron shells: K L M N
Electronic configuration: 1s^2 2s^2 2p^6 3s^2 3p^6 3d^8 4s^2
Valence electrons: 3d^8 4s^2
Crystal form: Cubic, face centered
Cross section σ: 4.54±0.10 barns
Vapor pressure: 2.37 × 10^2 Pa (at melting point)

Niobium

41

92.9064

Nióbio
Niobium
Niob
Niobio
ниобий
ניוביום

鈮 ニオブ

Naturally occurring isotope: 93
Density: 8.57 g/cm^3 (20°C)
Melting point: 2468±10°C   Boiling point: 4742°C
Latent heat of fusion: 288.4 J/g
Specific heat: 0.265 J/g/°C (25°C)
Coefficient of lineal thermal expansion: 7.31 × 10^{-6} cm/cm/°C (20°C)
Thermal conductivity: 0.537 w/cm/°C (25°C)
Electrical resistivity: 14.6 × 10^{-6} ohm-cm (20°C)
Ionization potential (1st): 6.88 eV
Electron work function ϕ: 4.3 eV
Oxidation potential: Nb → Nb^{3+} + 3ϵ = 1.099 V
Chemical valence: 2, 3, 4, *5*
Electrochemical equivalents: 0.69327 g/amp-hr
Ionic radius: 0.69 Å (Nb^{5+})
Valence electron potential ($-\epsilon$V): 104
Principal quantum number: 5
Principal electron shells: K L M N O
Electronic configuration: 1s^2 2s^2 2p^6 3s^2 3p^6 3d^{10} 4s^2 4p^6 4d^4 5s^1
Valence electrons: 4d^4 5s^1
Crystal form: Cubic, face centered
Cross section σ: 1.15±0.05 barns
Vapor pressure: 7.55 × 10^{-2} Pa (at melting point)

Nitrogen

VA

14.0067	N 7
30.97376	P 15
74.9216	As 33
121.75	Sb 51
208.9804	Bi 83

7

14.0067

Nitrogênio

Azote

Stickstoff

Nitrógeno

азот

חנקן

氮 窒素

Naturally occurring isotopes: 14, 15
Density: 1.165×10^{-3} g/cm^3 (20°C)
Melting point: -209.86°C   **Boiling point:** -195.8°C
Latent heat of fusion: 51.41 J/g (N$_2$)
Specific heat: 1.040 J/g/°C (N$_2$) (25°C)
Thermal conductivity: 0.2598 mw/cm/°C (27°C at 1 atm)
Ionization potential (1st): 14.534 eV
Oxidation potential: $N_2 + 2H_2O \rightarrow H_2N_2O_2 + 2H^+ + 2\epsilon = -2.65$ V
Chemical valence: $-3, 3, 5$
Electrochemical equivalents: 0.10452 g/amp-hr
Ionic radius: 0.13 Å (N^{5+})
Valence electron potential ($-\epsilon$V): 550
Principal quantum number: 2
Principal electron shells: K L
Electronic configuration: 1s^2 2s^2 2p^3
Valence electrons: 2s^2 2p^3
Crystal form: Hexagonal, close packed
Cross section σ: 1.9 barns

Nobelium

102

259

Actinide Series

Nobélio
Nobelium
Nobelium
Nobelio
нобелий
נובליום
ノーベリウム

232.03807	231.0359	238.029	237.0482	244.06423	243.0614	247.07038	247.07032	251.07961	254.08805
Th	Pa	U	Np	Pu	Am	Cm	Bk	Cf	Es
90	91	92	93	94	95	96	97	98	99
257.09515	258	259	260						
Fm	Md	No	Lr						
100	101	102	103						

Naturally occurring isotopes: None

Ionization potential (1st): 6.65 eV

Oxidation potential: $No \rightarrow No^{2+} + 2\epsilon = 2.5$ V

Chemical valence: *2*, 3

Electrochemical equivalents: 4.83 g/amp-hr

Ionic radius: 1.1 Å (est) (No^{2+})

Valence electron potential ($-\epsilon V$): (26)

Principal quantum number: 7

Principal electron shells: K L M N O P Q

Electronic configuration: $1s^2\ 2s^2\ 2p^6\ 3s^2\ 3p^6\ 3d^{10}\ 4s^2\ 4p^6\ 4d^{10}\ 4f^{14}\ 5s^2\ 5p^6$
$5d^{10}\ 5f^{14}\ 6s^2\ 6p^6\ 7s^2$

Valence electrons: $5f^{14}\ 7s^2$

Half life: ~59 minutes

Osmium

76

190.2

VIII		
55.847 Fe 26	58.9332 Co 27	58.70 Ni 28
101.07 Ru 44	102.9055 Rh 45	106.4 Pd 46
190.2 Os 76	192.22 Ir 77	195.09 Pt 78
	109	

Osmio

Osmium

Osmium

Osmio

осмий

אוסמיום

銥 オスミウム

Naturally occurring isotopes: 192, 190, 189, 188, 187, 186, 184

Density: 22.61 g/cm^3 (20°C)

Melting point: 3045 ± 30°C   **Boiling point:** 5027 ± 100°C

Latent heat of fusion: 154.1 J/g

Specific heat: 0.13 J/g/°C (25°C)

Coefficient of lineal thermal expansion: 6.3 × 10^{-6} cm/cm/°C (20°C)

Thermal conductivity: 0.876 w/cm/°C (25°C)

Electrical resistivity: 9.5 × 10^{-6} ohm-cm (20°C)

Ionization potential (1st): 8.7 eV

Electron work function ϕ: 4.83 eV

Oxidation potential: $Os + 4H_2O \rightarrow OsO_4 + 8H^+ + 8\epsilon = -0.85$ V

Chemical valence: 0, 1, 2, 3, *4*, 5, 6, 7, 8

Electrochemical equivalents: 1.774 g/amp-hr

Ionic radius: 0.630 Å (Os^{4+})

Valence electron potential (−εV): 91.4

Principal quantum number: 6

Principal electron shells: K L M N O P

Electronic configuration: $1s^2\ 2s^2\ 2p^6\ 3s^2\ 3p^6\ 3d^{10}\ 4s^2\ 4p^6\ 4d^{10}\ 4f^{14}\ 5s^2\ 5p^6$
    $5d^6\ 6s^2$

Valence electrons: $5d^6\ 6s^2$

Crystal form: Hexagonal, close packed

Cross section σ: 15.3 ± 0.7 barns

Vapor pressure: 2.52 Pa (at melting point)

Oxygen

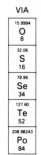

VIA

15.9994	O 8
32.06	S 16
78.96	Se 34
127.60	Te 52
208.98243	Po 84

Oxigênio
Oxygène
Sauerstoff
Oxigeno
кислород
חמצן
氧 酸素

Naturally occurring isotopes: 16, 18, 17
Density: 1.429×10^{-3} g/cm³ (0°C)
Melting point: -218.4°C   **Boiling point:** -182.962°C
Latent heat of fusion: 26.17 J/g (O_2)
Specific heat: 0.9174 J/g/°C (O_2) (25°C)
Thermal conductivity: 0.2674 w/cm/°C (25°C at 1 atm)
Ionization potential (1st): 13.618 eV
Oxidation potential: $2H_2O$ (liquid) $\rightarrow O_2 + 4H^+ + 4\epsilon = -1.229$ V
Chemical valence: -2
Electrochemical equivalents: 0.29847 g/amp-hr
Ionic radius: 1.40 Å (O^{2-})
Valence electron potential ($-\epsilon V$): -20.6
Principal quantum number: 2
Principal electron shells: K L
Electronic configuration: $1s^2\ 2s^2\ 2p^4$
Valence electrons: $2s^2\ 2p^4$
Crystal form: Cubic
Cross section σ: 0.27 mbarns

Palladium

46

106.4

VIII		
55.847 Fe 26	58.9332 Co 27	58.70 Ni 28
101.07 Ru 44	102.9055 Rh 45	106.4 Pd 46
190.2 Os 76	192.22 Ir 77	195.09 Pt 78
	109	

Paládio
Palladium
Palladium
Paladio
палладий
סלדיום

Naturally occurring isotopes: 106, 108, 105, 110, 104, 102
Density: 12.023 g/cm^3 (20°C)
Melting point: 1554°C   **Boiling point:** 3140°C
Latent heat of fusion: 157.4 J/g
Specific heat: 0.244 J/g/°C (25°C)
Coefficient of lineal thermal expansion: 11.67 × 10^{-6} cm/cm/°C (0°C)
Thermal conductivity: 0.718 w/cm/°C (25°C)
Electrical resistivity: 10.54 × 10^{-6} ohm-cm (20°C)
Ionization potential (1st): 8.34 eV
Electron work function ϕ: 5.12 eV
Oxidation potential: $Pd \rightarrow Pd^{2+} + 2\epsilon = -0.987$ V
Chemical valence: *2*, 3, 4
Electrochemical equivalents: 1.985 g/amp-hr
Ionic radius: 0.86 Å (Pd^{2+})
Valence electron potential ($-\epsilon V$): 33
Principal quantum number: 5
Principal electron shells: K L M N O
Electronic configuration: $1s^2\ 2s^2\ 2p^6\ 3s^2\ 3p^6\ 3d^{10}\ 4s^2\ 4p^6\ 4d^{10}$
Valence electrons: $4d^{10}$
Crystal form: Cubic, face centered
Cross section σ: 6.0 ± 1.0 barns
Vapor pressure: 1.33 Pa (at melting point)

Phosphorus

15

30.97376

VA

14.0067 N 7
30.97376 P 15
74.9216 As 33
121.75 Sb 51
208.9804 Bi 83

Fósforo

Phosphore

Phosphor

Fósforo

фосфор

זרחן

磷 燐

Naturally occurring isotope: 31
Density: 1.828 g/cm³ (white), 2.34 g/cm³ (red), 2.699 g/cm³ (black) (all at 20°C)
Melting point: 44.1°C (white)   **Boiling point:** 280.3°C (white)
Latent heat of fusion: 20.28 J/g (white)
Specific heat: 0.7697 J/g/°C (white) (25°C)
Coefficient of lineal thermal expansion: 125×10^{-6} cm/cm/°C (25°C)
Thermal conductivity: 2.36 mw/cm/°C (white) (25°C)
Electrical resistivity: 10^{11} ohm-cm (white) (20°C)
Ionization potential (1st): 10.486 eV
Oxidation potential: $P + 2H_2O \rightarrow H_3PO_2 + H^+ + \epsilon = 0.508$ V
Chemical valence: $-3, 3, 5$
Electrochemical equivalents: 0.23113 g/amp-hr
Ionic radius: 0.38 Å (P^{5+})
Valence electron potential ($-\epsilon V$): 190
Principal quantum number: 3
Principal electron shells: K L M
Electronic configuration: $1s^2\ 2s^2\ 2p^6\ 3s^2\ 3p^3$
Valence electrons: $3s^2\ 3p^3$
Crystal form: Cubic
Cross section σ: 0.19 barns
Vapor pressure: 20.8 Pa (at melting point)

Four allotropes of phosphorus have different melting points, crystal forms, colors, and electrical conductivities. The black variety has the highest electrical conductivity.

Platinum

78

195.09

VIII		
55.847 Fe 26	58.9332 Co 27	58.70 Ni 28
101.07 Ru 44	102.9055 Rh 45	106.4 Pd 46
190.2 Os 76	192.22 Ir 77	195.09 Pt 78
	109	

Platina

Platine

Plátin

Platino

платина

פלטין

鉑 白金〔プラチナ〕

Naturally occurring isotopes: 195, 194, 196, 198, 192, 190
Density: 21.45 g/cm^3 (20°C)
Melting point: 1773.5°C   **Boiling point:** 3827 ± 100°C
Latent heat of fusion: 100.9 J/g
Specific heat: 0.133 J/g/°C (25°C)
Coefficient of lineal thermal expansion: 9.5 × 10^{-6} cm/cm/°C (25°C)
Thermal conductivity: 0.716 w/cm/°C (25°C)
Electrical resistivity: 9.85 × 10^{-6} ohm-cm (0°C)
Ionization potential (1st): 8.96 eV
Electron work function ϕ: 5.65 eV
Oxidation potential: Pt → Pt^{2+} + 2ϵ = −1.2 V
Chemical valence: 2, 3, *4*
Electrochemical equivalents: 1.8197 g/amp-hr
Ionic radius: 0.625 Å (Pt^{4+})
Valence electron potential ($-\epsilon$V): 92.2
Principal quantum number: 6
Principal electron shells: K L M N O P
Electronic configuration: 1s^2 2s^2 2p^6 3s^2 3p^6 3d^{10} 4s^2 4p^6 4d^{10} 4f^{14} 5s^2 5p^6
     5d^9 6s^1
Valence electrons: 5d^9 6s^1
Crystal form: Cubic, face centered
Cross section σ: 9 ± 1 barns
Vapor pressure: 3.12 × 10^{-2} Pa (at melting point)

Plutonium

Actinide Series

232.03807	231.0359	238.029	237.0482	244.06423	243.0614	247.07038	247.07032	251.07961	254.08805
Th	Pa	U	Np	Pu	Am	Cm	Bk	Cf	Es
90	91	92	93	94	95	96	97	98	99
257.09515	258	259	260						
Fm	Md	No	Lr						
100	101	102	103						

Plutônio
Plutonium
Plutonium
Plutonio
плутоний
סלוטוניום

鈈 プリトニウム

Naturally occurring isotope: 242 (trace)
Density: 19.737 g/cm^3 (25°C)
Melting point: 639.5°C   **Boiling point:** 3232°C
Latent heat of fusion: 11 J/g
Specific heat: 0.14 J/g/°C (25°C)
Coefficient of lineal thermal expansion: 42.3 × 10^{-6} cm/cm/°C (20°C)
Thermal conductivity: 0.0670 w/cm/°C (25°C)
Electrical resistivity: 146.45 × 10^{-6} ohm-cm (0°C)
Ionization potential (1st): 6.06 eV
Oxidation potential: Pu → Pu^{3+} + 3ε = 2.031 V
Chemical valence: 3, *4*, 5, 6, 7
Electrochemical equivalents: 2.2765 g/amp-hr
Ionic radius: 0.887 Å (Pu^{4+})
Valence electron potential (−εV): 64.9
Principal quantum number: 7
Principal electron shells: K L M N O P Q
Electronic configuration: 1s^2 2s^2 2p^6 3s^2 3p^6 3d^{10} 4s^2 4p^6 4d^{10} 4f^{14} 5s^2 5p^6
        5d^{10} 5f^6 6s^2 6p^6 7s^2
Valence electrons: 5f^6 7s^2
Crystal form: Monoclinic
Half life: 8.3 × 10^7 years
Cross section σ: 1.8 ± 0.3 barns

Polonium

84

208.98243

VIA

| 15.9994 O 8 |
| 32.06 S 16 |
| 78.96 Se 34 |
| 127.60 Te 52 |
| 208.98243 Po 84 |

Polônio
Polonium
Polonium
Polonio
полоний
סולוניום
鈽 ポロニウム

Naturally occurring isotopes: None
Density: 9.20 g/cm^3 (20°C)
Melting point: 254°C   **Boiling point:** 962°C
Latent heat of fusion: 60.1 J/g
Specific heat: 0.13 J/g/°C (25°C)
Coefficient of lineal thermal expansion: 23.5 × 10^{-6} cm/cm/°C (20°C)
Electrical resistivity: 42 × 10^{-6} ohm-cm (0°C)
Ionization potential (1st): 8.42 eV
Oxidation potential: Po → Po^{2+} + 2ϵ = −0.65 V
Chemical valence: −2, 0, 2, 4, 6
Electrochemical equivalents: 3.8986 g/amp-hr
Ionic radius: 2.30 Å (Po^{2-})
Valence electron potential (−ϵV): −12.5
Principal quantum number: 6
Principal electron shells: K L M N O P
Electronic configuration: 1s^2 2s^2 2p^6 3s^2 3p^6 3d^{10} 4s^2 4p^6 4d^{10} 4f^{14} 5s^2 5p^6
    5d^{10} 6s^2 6p^4
Valence electrons: 6s^2 6p^4
Crystal form: Cubic, body centered
Half life: 103 years
Vapor pressure: 1.76 × 10^{-2} Pa (at melting point)

Potassium

19

39.098

IA
1.0079 **H** 1
6.941 **Li** 3
22.98977 **Na** 11
39.098 **K** 19
85.4678 **Rb** 37
132.9054 **Cs** 55
223.01976 **Fr** 87

Potássio
Potassium
Kalium
Potasio
калий
אשלגן
鉀 カリウム

Naturally occurring isotopes: 39, 41, 40
Density: 0.862 g/cm^3 (20°C)
Melting point: 63.25°C   Boiling point: 759.9°C
Latent heat of fusion: 59.33 J/g
Specific heat: 0.757 J/g/°C (25°C)
Coefficient of lineal thermal expansion: 83 × 10^{-6} cm/cm/°C (20°C)
Thermal conductivity: 1.025 w/cm/°C (25°C)
Electrical resistivity: 7.20 × 10^{-6} ohm-cm (20°C)
Ionization potential (1st): 4.341 eV
Electron work function φ: 2.30 eV
Oxidation potential: K → K$^+$ + ϵ = 2.925 V
Chemical valence: 1
Electrochemical equivalents: 1.4587 g/amp-hr
Ionic radius: 1.38 Å (K$^+$)
Valence electron potential (−εV): 10.4
Principal quantum number: 4
Principal electron shells: K L M N
Electronic configuration: 1s^2 2s^2 2p^6 3s^2 3p^6 4s^1
Valence electrons: 4s^1
Crystal form: Cubic, body centered
Cross section σ: 2.1 barns
Vapor pressure: 1.06 × 10^{-4} Pa (at melting point)

Praseodymium

59

140.9077

Lanthanide Series

Praséodímio

Praséodyne

Praseodym

Praseodimio

празеодимий

פרסיאודים

鐠 プラセオジム

140.12	140.9077	144.24	144.913	150.4	151.96	157.25	158.9254	162.50	164.9304
Ce 58	Pr 59	Nd 60	Pm 61	Sm 62	Eu 63	Gd 64	Tb 65	Dy 66	Ho 67
167.26	168.9342	173.04	174.97						
Er 68	Tm 69	Yb 70	Lu 71						

Naturally occurring isotope: 141
Density: 6.773 g/cm^3 (25°C)
Melting point: 931°C   **Boiling point:** 3512°C
Latent heat of fusion: 71.3 J/g
Specific heat: 0.193 J/g/°C (25°C)
Coefficient of lineal thermal expansion: 6.5 × 10^{-6} cm/cm/°C (25°C)
Thermal conductivity: 0.125 w/cm/°C (25°C)
Electrical resistivity: 68 × 10^{-6} ohm-cm (25°C)
Ionization potential (1st): 5.42 eV
Electron work function ϕ: 2.7 eV
Oxidation potential: Pr → Pr^{3+} + 3ϵ = 2.462 V
Chemical valence: *3,* 4
Electrochemical equivalents: 1.7524 g/amp-hr
Ionic radius: 1.013 Å (Pr^{3+})
Valence electron potential ($-\epsilon$V): 42.64
Principal quantum number: 6
Principal electron shells: K L M N O P
Electronic configuration: 1s^2 2s^2 2p^6 3s^2 3p^6 3d^{10} 4s^2 4p^6 4d^{10} 4f^3 5s^2
    5p^6 6s^2
Valence electrons: 4f^3 6s^2
Crystal form: Hexagonal, close packed
Cross section σ: 3.9±0.5 barns

 **Promethium**

61

144.913

Lanthanide Series

140.12	140.9077	144.24	144.913	150.4	151.96	157.25	158.9254	162.50	164.9304
Ce	Pr	Nd	Pm	Sm	Eu	Gd	Tb	Dy	Ho
58	59	60	61	62	63	64	65	66	67
167.26	168.9342	173.04	174.97						
Er	Tm	Yb	Lu						
68	69	70	71						

Promécio
Prometheum
Prometheum
Promecio
прометий
פרומתיום

鉅 プロメチウム

Naturally occurring isotopes: None
Density: 7.22 ± 0.02 g/cm³ (25°C)
Melting point: 1168 ± 6°C   **Boiling point:** 2460°C
Latent heat of fusion: 86.7 J/g
Specific heat: 0.185 J/g/°C (25°C)
Thermal conductivity: 0.179 w/cm/°C (25°C)
Ionization potential (1st): 5.55 eV
Oxidation potential: $Pm \rightarrow Pm^{3+} + 3\epsilon = 2.423$ V
Chemical valence: 3
Electrochemical equivalents: 1.8022 g/amp-hr
Ionic radius: 0.979 Å (Pm^{3+})
Valence electron potential ($-\epsilon V$): 44.1
Principal quantum number: 6
Principal electron shells: K L M N O P
Electronic configuration: $1s^2\ 2s^2\ 2p^6\ 3s^2\ 3p^6\ 3d^{10}\ 4s^2\ 4p^6\ 4d^{10}\ 4f^5\ 5s^2$
    $5p^6\ 6s^2$
Valence electrons: $4f^5\ 6s^2$
Crystal form: Hexagonal
Half life: 17.7 years

Protactinium

91

231.0359

Actinide Series

232.03807	231.0359	238.029	237.0482	244.06423	243.0614	247.07038	247.07032	251.07961	254.08805
Th 90	Pa 91	U 92	Np 93	Pu 94	Am 95	Cm 96	Bk 97	Cf 98	Es 99
257.09515	258	259	260						
Fm 100	Md 101	No 102	Lr 103						

Naturally occurring isotope: 231 (minute quantities only)
Density: 15.37 g/cm^3 (25°C)
Melting point: 1575°C
Latent heat of fusion: 65 J/g
Specific heat: 0.12 J/g/°C (25°C)
Coefficient of lineal thermal expansion: 11.2 × 10^{-6} cm/cm/°C (25°C)
Ionization potential (1st): 5.89 eV
Chemical valence: 3, 4, *5*
Electrochemical equivalents: 1.7240 g/amp-hr
Oxidation potential: Pa → Pa^{3+} + 3ϵ = 1.6 V
Ionic radius: 0.78 Å (Pa^{5+})
Valence electron potential ($-\epsilon$V): 92
Principal quantum number: 7
Principal electron shells: K L M N O P Q
Electronic configuration: 1s^2 2s^2 2p^6 3s^2 3p^6 3d^{10} 4s^2 4p^6 4d^{10} 4f^{14} 5s^2 5p^6
    5d^{10} 5f^2 6s^2 6p^6 6d^1 7s^2
Valence electrons: 5f^2 6d^1 7s^2
Crystal form: Tetragonal
Half life: 3.248 × 10^4 years
Cross section σ: 200 ± 10 barns

Radium

88

226.02544

IIA

IIA
9.01218 Be 4
24.305 Mg 12
40.08 Ca 20
87.62 Sr 38
137.34 Ba 56
226.02544 Ra 88

Rádio
Radium
Radium
Radio
радий
רדיום

鐳 ラジウム

Naturally occurring isotope: 226 (minute quantities only)
Density: 5.5 g/cm^3 (extrapolated) (20°C)
Melting point: 700°C   **Boiling point:** 1140°C
Latent heat of fusion: 37 J/g (est)
Specific heat: 0.120 J/g/°C (25°C)
Thermal conductivity: 0.186 w/cm/°C (20°C)
Ionization potential (1st): 5.279 eV
Oxidation potential: $Ra \rightarrow Ra^{2+} + 2\epsilon = 2.916$ V
Chemical valence: 2
Electrochemical equivalents: 4.2165 g/amp-hr
Ionic radius: 1.43 Å (Ra^{2+})
Valence electron potential ($-\epsilon V$): 20.1
Principal quantum number: 7
Principal electron shells: K L M N O P Q
Electronic configuration: $1s^2\ 2s^2\ 2p^6\ 3s^2\ 3p^6\ 3d^{10}\ 4s^2\ 4p^6\ 4d^{10}\ 4f^{14}\ 5s^2\ 5p^6$ $5d^{10}\ 6s^2\ 6p^6\ 7s^2$
Valence electrons: $7s^2$
Half life: 1622 years
Cross section σ: 20 ± 3 barns
Vapor pressure: 3.27×10^2 Pa (at melting point)

Radon

86

222.01761

O

4.00260
He
2
20.179
Ne
10
39.948
Ar
18
83.80
Kr
36
131.30
Xe
54
222.01761
Rn
86

Radônio

Radon

Radon

Radòn

радон

רדון

Naturally occurring isotopes: None (radium decay product)
Density: 9.96×10^{-3} g/cm^3 (20°C)
Melting point: -71°C   **Boiling point:** -61.8°C
Latent heat of fusion: 13.1 J/g
Specific heat: 0.09362 J/g/°C (25°C)
Thermal conductivity: 0.0364 mw/cm/°C (27°C)
Ionization potential (1st): 10.748 eV
Chemical valence: 0
Principal quantum number: 6
Principal electron shells: K L M N O P
Electronic configuration: $1s^2\ 2s^2\ 2p^6\ 3s^2\ 3p^6\ 3d^{10}\ 4s^2\ 4p^6\ 4d^{10}\ 4f^{14}\ 5s^2\ 5p^6$
    $5d^{10}\ 6s^2\ 6p^6$
Valence electrons: ($6s^2\ 6p^6$)
Crystal form: Cubic, face centered
Half life: 3.824 days
Cross section σ: 0.72 ± 0.07 barns

Rhenium

VIIB
54.9380 **Mn** 25
96.906 **Tc** 43
186.2 **Re** 75
107

75

186.2

Rênio

Rhenium

Rhenium

Renio

рений

רניום

鍊 レニウム

Naturally occurring isotopes: 187, 185
Density: 21.04 g/cm³ (20°C)
Melting point: 3180°C   **Boiling point:** 5627°C (est)
Latent heat of fusion: 177.6 J/g
Specific heat: 0.137 J/g/°C (25°C)
Coefficient of lineal thermal expansion: 6.7×10^{-6} cm/cm/°C (25°C)
Thermal conductivity: 0.480 w/cm/°C (25°C)
Electrical resistivity: 19.3×10^{-6} ohm-cm (20°C)
Ionization potential (1st): 7.88 eV
Electron work function ϕ: 4.96 eV
Oxidation potential: $Re + 2H_2O \rightarrow ReO_2 + 4H^+ + 4\epsilon = -0.2513$ V
Chemical valence: 0, 1, 2, 3, 4, 5, 6, 7
Electrochemical equivalents: 0.9924 g/amp-hr
Ionic radius: 0.56 Å (Re^{7+})
Valence electron potential ($-\epsilon$V): 180
Principal quantum number: 6
Principal electron shells: K L M N O P
Electronic configuration: $1s^2\ 2s^2\ 2p^6\ 3s^2\ 3p^6\ 3d^{10}\ 4s^2\ 4p^6\ 4d^{10}\ 4f^{14}\ 5s^2\ 5p^6$
    $5d^5\ 6s^2$
Valence electrons: $5d^5\ 6s^2$
Crystal form: Hexagonal, close packed
Cross section σ: 85 ± 5 barns
Vapor pressure: 3.24 Pa (at melting point)

Rh **Rhodium**

45

102.9055

VIII		
55.847 Fe 26	58.9332 Co 27	58.70 Ni 28
101.07 Ru 44	102.9055 Rh 45	106.4 Pd 46
190.2 Os 76	192.22 Ir 77	195.09 Pt 78
	109	

Ródio
Rhodium
Rhodium
Rodio
родий
רודיום

銠 ロジウム

Naturally occurring isotope: 103
Density: 12.41 g/cm^3 (20°C)
Melting point: 1966 ± 3°C   **Boiling point:** 3727 ± 100°C
Latent heat of fusion: 211.6 J/g
Specific heat: 0.24 J/g/°C (25°C)
Coefficient of lineal thermal expansion: 8.3 × 10^{-6} cm/cm/°C (20°C)
Thermal conductivity: 1.50 w/cm/°C (25°C)
Electrical resistivity: 4.51 × 10^{-6} ohm-cm (20°C)
Ionization potential (1st): 7.46 eV
Electron work function ϕ: 4.98 eV
Oxidation potential: Rh → Rh^{3+} + 3ϵ = −0.80 V
Chemical valence: 2, *3*, 4, 5, 6
Electrochemical equivalents: 1.2798 g/amp-hr
Ionic radius: 0.68 Å (Rh^{3+})
Valence electron potential (−ϵV): 64
Principal quantum number: 5
Principal electron shells: K L M N O
Electronic configuration: 1s^2 2s^2 2p^6 3s^2 3p^6 3d^{10} 4s^2 4p^6 4d^8 5s^1
Valence electrons: 4d^8 5s^1
Crystal form: Cubic, face centered
Cross section σ: 150 ± 5 barns
Vapor pressure: 6.33 × 10^{-1} Pa (at melting point)

Rubidium

37

85.4678

IA
1.0079 **H** 1
6.941 **Li** 3
22.98977 **Na** 11
39.098 **K** 19
85.4678 **Rb** 37
132.9054 **Cs** 55
223.01976 **Fr** 87

Rubidio
Rubidium
Rubidium
Rubidio
рубидий

רוֹבּידיוֹם

鉫 ルビジウム

Naturally occurring isotopes: 85, 87
Density: 1.532 g/cm³ (20°C)
Melting point: 38.89°C   **Boiling point:** 686°C
Latent heat of fusion: 27.43 J/g
Specific heat: 0.3634 J/g/°C (25°C)
Coefficient of lineal thermal expansion: 90×10^{-6} cm/cm/°C (25°C)
Thermal conductivity: 0.582 w/cm/°C (25°C)
Electrical resistivity: 12.84×10^{-6} ohm-cm (20°C)
Ionization potential (1st): 4.177 eV
Electron work function ϕ: 2.16 eV
Oxidation potential: $Rb \rightarrow Rb^{+} + \epsilon = 2.925$ V
Chemical valence: 1
Electrochemical equivalents: 3.1888 g/amp-hr
Ionic radius: 1.52 Å (Rb^{+})
Valence electron potential ($-\epsilon V$): 9.47
Principal quantum number: 5
Principal electron shells: K L M N O
Electronic configuration: $1s^2\ 2s^2\ 2p^6\ 3s^2\ 3p^6\ 3d^{10}\ 4s^2\ 4p^6\ 5s^1$
Valence electrons: $5s^1$
Crystal form: Cubic, body centered
Cross section σ: 0.5 ± 0.1 barns
Vapor pressure: 1.56×10^{-4} Pa (at melting point)

Ruthenium

44

101.07

VIII		
55.847 Fe 26	58.9332 Co 27	58.70 Ni 28
101.07 Ru 44	102.9055 Rh 45	106.4 Pd 46
190.2 Os 76	192.22 Ir 77	195.09 Pt 78
	109	

Rutênio
Ruthénium
Ruthenium
Rutenio
рутений
רותניום

釕 ルテニウム

Naturally occurring isotopes: 102, 104, 101, 99, 100, 96, 98
Density: 12.45 g/cm^3 (20°C)
Melting point: 2310°C   **Boiling point:** 3900°C
Latent heat of fusion: 252.7 J/g
Specific heat: 0.238 J/g/°C (25°C)
Coefficient of lineal thermal expansion: 9.91 × 10^{-6} cm/cm/°C (50°C)
Thermal conductivity: 1.17 w/cm/°C (25°C)
Electrical resistivity: 6.80 × 10^{-6} ohm-cm (0°C)
Ionization potential (1st): 7.37 eV
Electron work function ϕ: 4.71 eV
Oxidation potential: $Ru + 5Cl^- \rightarrow RuCl_3{}^{2-} + 3\epsilon = -0.601$ V
Chemical valence: 1, 2, *3*, 4, 5, 6, 7, 8
Electrochemical equivalents: 1.2570 g/amp-hr
Ionic radius: 0.68 Å (Ru^{4+})
Valence electron potential ($-\epsilon$V): 64
Principal quantum number: 5
Principal electron shells: K L M N O
Electronic configuration: $1s^2\ 2s^2\ 2p^6\ 3s^2\ 3p^6\ 3d^{10}\ 4s^2\ 4p^6\ 4d^7\ 5s^1$
Valence electrons: $4d^7\ 5s^1$
Crystal form: Hexagonal, close packed
Cross section σ: 3.0 ± 0.8 barns
Vapor pressure: 1.40 Pa (at melting point)

Sm Samarium

62

150.4

Lanthanide Series

140.12	140.9077	144.24	144.913	150.4	151.96	157.25	158.9254	162.50	164.9304
Ce	Pr	Nd	Pm	Sm	Eu	Gd	Tb	Dy	Ho
58	59	60	61	62	63	64	65	66	67
167.26	168.9342	173.04	174.97						
Er	Tm	Yb	Lu						
68	69	70	71						

Samário
Samarium
Samarium
Samario
самарий
סמריום

 サマリウム

Naturally occurring isotopes: 152, 154, 147, 149, 148, 150, 144
Density: 7.520 g/cm^3 (25°C)
Melting point: 1077°C   **Boiling point:** 1791°C
Latent heat of fusion: 73.8 J/g
Specific heat: 0.196 J/g/°C (25°C)
Thermal conductivity: 0.133 w/cm/°C (25°C)
Electrical resistivity: 88 × 10^{-6} ohm-cm (25°C)
Ionization potential (1st): 5.63 eV
Electron work function φ: 2.7 eV
Oxidation potential: Sm → Sm^{3+} + 3ε = 2.414 V
Chemical valence: 2, *3*
Electrochemical equivalents: 1.870 g/amp-hr
Ionic radius: 0.964 Å (Sm^{3+})
Valence electron potential (−εV): 44.8
Principal quantum number: 6
Principal electron shells: K L M N O P
Electronic configuration: 1s^2 2s^2 2p^6 3s^2 3p^6 3d^{10} 4s^2 4p^6 4d^{10} 4f^6 5s^2
    5p^6 6s^2
Valence electrons: 4f^6 6s^2
Crystal form: Rhombohedral
Cross section σ: 5820 ± 100 barns
Vapor pressure: 5.63 × 10^2 Pa (at melting point)

Scandium

21

44.95592

IIIB
44.95592 **Sc** 21
88.9059 **Y** 39
138.9055 **La** 57
227.02777 **Ac** 89

Escândio
Scandium
Scandium
Escandio
скандий
סקנדיום
钪　スカンジウム

Naturally occurring isotope: 45
Density: 2.989 g/cm^3 (25°C)
Melting point: 1541°C   **Boiling point:** 2831°C
Latent heat of fusion: 358.6 J/g
Specific heat: 0.568 J/g/°C (25°C)
Coefficient of lineal thermal expansion: 12×10^{-6} cm/cm/°C (25°C)
Thermal conductivity: 0.158 w/cm/°C (25°C)
Electrical resistivity: 61.0×10^{-6} ohm-cm (20°C)
Ionization potential (1st): 6.54 eV
Electron work function ϕ**:** 3.5 eV
Oxidation potential: $Sc \rightarrow Sc^{3+} + 3\epsilon = 2.077$ V
Chemical valence: 3
Electrochemical equivalents: 0.55914 g/amp-hr
Ionic radius: 0.745 Å (Sc^{3+})
Valence electron potential $(-\epsilon V)$**:** 58.0
Principal quantum number: 4
Principal electron shells: K L M N
Electronic configuration: $1s^2\ 2s^2\ 2p^6\ 3s^2\ 3p^6\ 3d^1\ 4s^2$
Valence electrons: $3d^1\ 4s^2$
Crystal form: Hexagonal, close packed
Cross section σ**:** 25 ± 2 barns
Vapor pressure: 2.21×10 Pa (at melting point)

Se **Selenium**

34

78.96

VIA
15.9994 **O** 8
32.06 **S** 16
78.96 **Se** 34
127.60 **Te** 52
208.98243 **Po** 84

Selênio

Sélénium

Selen

Selenio

селен

סלן

硒 セレン

Naturally occurring isotopes: 80, 78, 82, 76, 77, 74
Density: 4.792 g/cm³ (gray) (20°C)
Melting point: 217°C (gray)   **Boiling point:** 684.9 ± 1.0°C
Latent heat of fusion: 68.93 J/g
Specific heat: 0.1606 J/g/°C (Se₂) (25°C)
Coefficient of lineal thermal expansion: 36.8 cm/cm/°C (20°C)
Thermal conductivity: 0.0452 w/cm/°C (along C-axis at 25°C)
Electrical resistivity: 1 ohm-cm (20°C)
Ionization potential (1st): 9.752 eV
Electron work function φ: 5.9 eV
Oxidation potential: $Se + 3H_2O \rightarrow H_2SeO_3 + 4H^+ + 4\epsilon = -0.740$ V
Chemical valence: −2, *4*, 6
Electrochemical equivalents: 0.73650 g/amp-hr
Ionic radius: 0.50 Å (Se⁴⁺)
Valence electron potential (−εV): 120
Principal quantum number: 4
Principal electron shells: K L M N
Electronic configuration: $1s^2\ 2s^2\ 2p^6\ 3s^2\ 3p^6\ 3d^{10}\ 4s^2\ 4p^4$
Valence electrons: $4s^2\ 4p^4$
Crystal forms: Hexagonal, monoclinic, amorphous
Cross section σ: 12.2 ± 0.6 barns
Vapor pressure: 6.95×10^{-1} Pa (at melting point)

Si **Silicon**

14

28.0855

IVA

| 12.011 |
| C |
| 6 |

| 28.0855 |
| Si |
| 14 |

| 72.59 |
| Ge |
| 32 |

| 118.69 |
| Sn |
| 50 |

| 207.2 |
| Pb |
| 82 |

Silício

Silicium

Silizium

Silicio

кремний

צורן

硅 珪素

Naturally occurring isotopes: 28, 29, 30

Density: 2.329 g/cm^3 (25°C)

Melting point: 1410°C   **Boiling point:** 2355°C

Latent heat of fusion: 1.655 J/g

Specific heat: 0.712 J/g/°C (25°C)

Coefficient of lineal thermal expansion: 4.2 × 10^{-6} cm/cm/°C (25°C)

Thermal conductivity: 1.49 w/cm/°C (25°C)

Electrical resistivity: 3.5 ohm-cm (20°C)

Ionization potential (1st): 8.151 eV

Electron work function φ: 4.52 eV

Oxidation potential: Si + 2H$_2$O → SiO$_2$ + 4H$^+$ + 4ε = 0.857 V

Chemical valence: −4, −1, 1, *4*

Electrochemical equivalents: 0.26197 g/amp-hr

Ionic radius: 0.400 Å (Si^{4+})

Valence electron potential (− εV): 144

Principal quantum number: 3

Principal electron shells: K L M

Electronic configuration: 1s^2 2s^2 2p^6 3s^2 3p^2

Valence electrons: 3s^2 3p^2

Crystal form: Cubic, diamond

Cross section σ: 160 ± 20 mbarns

Vapor pressure: 4.77 Pa (at melting point)

Silver

		IB	
47		63.546 **Cu** 29	**Prata**
		107.868 **Ag** 47	**Argent**
107.868		196.9665 **Au** 79	**Silber**

Prata
Argent
Silber
Plata
серебро
כֶסֶף

銀 銀

Naturally occurring isotopes: 107, 109
Density: 10.50 g/cm^3 (20°C)
Melting point: 961.93°C   **Boiling point:** 2212°C
Latent heat of fusion: 104.8 J/g
Specific heat: 0.2350 J/g/°C (25°C)
Coefficient of lineal thermal expansion: 18.62 $\times$ 10^{-6} cm/cm/°C (17°C)
Thermal conductivity: 4.29 w/cm/°C (25°C)
Electrical resistivity: 1.586 $\times$ 10^{-6} ohm-cm (20°C)
Ionization potential (1st): 7.576 eV
Electron work function ϕ: 4.26 eV
Oxidation potentials: Ag $\rightarrow$ Ag$^+$ + ϵ = -0.7991 V
$\quad\quad\quad\quad\quad\quad\quad$ Ag$^+$ $\rightarrow$ Ag^{2+} + ϵ = -1.980 V
Chemical valence: *1*, 2, 3
Electrochemical equivalents: 4.0246 g/amp-hr
Ionic radius: 1.26 Å (Ag$^+$)
Valence electron potential ($-\epsilon$V): 11.4
Principal quantum number: 5
Principal electron shells: K L M N O
Electronic configuration: 1s^2 2s^2 2p^6 3s^2 3p^6 3d^{10} 4s^2 4p^6 4d^{10} 5s^1
Valence electrons: (4d^{10}) 5s^1
Crystal form: Cubic, face centered
Cross section σ: 63.8$\pm$0.6 barns
Vapor pressure: 3.42 $\times$ 10^{-1} Pa (at melting point)

Sodium

IA
1.0079 **H** 1
6.941 **Li** 3
22.98977 **Na** 11
39.098 **K** 19
85.4678 **Rb** 37
132.9054 **Cs** 55
223.01976 **Fr** 87

11

22.98977

Sódio
Sodium
Natrium
Sodio
натрий
נתרן

鈉 ナトリウム

Naturally occurring isotopes: 23
Density: 0.9712 g/cm^3 (20°C)
Melting point: 97.81 ± 0.03°C   **Boiling point:** 882.9°C
Latent heat of fusion: 113 J/g
Specific heat: 1.23 J/g/°C (25°C)
Coefficient of lineal thermal expansion: 72 × 10^{-6} cm/cm/°C (25°C)
Thermal conductivity: 1.42 w/cm/°C (25°C)
Electrical resistivity: 4.33 × 10^{-6} ohm-cm (0°C)
Ionization potential (1st): 5.139 eV
Electron work function ϕ: 2.75 eV
Oxidation potential: Na $\rightarrow$ Na$^+$ + ϵ = 2.714 V
Chemical valence: 1
Electrochemical equivalents: 0.85775 g/amp-hr
Ionic radius: 1.02 Å (Na$^+$)
Valence electron potential ($-\epsilon$V): 14.1
Principal quantum number: 3
Principal electron shells: K L M
Electronic configuration: 1s^2 2s^2 2p^6 3s^1
Valence electrons: 3s^1
Crystal form: Cubic, body centered
Cross section σ: 534 ± 5 mbarns
Vapor pressure: 1.43 × 10^{-5} Pa (at melting point)

Sr Strontium

38

87.62

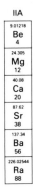

IIA

9.01218	
Be	
4	
24.305	
Mg	
12	
40.08	
Ca	
20	
87.62	
Sr	
38	
137.34	
Ba	
56	
226.02544	
Ra	
88	

Estrôncio
Strontium
Strontium
Estroncio
стронций
סטרונציום
鍶 ストロンチウム

Naturally occurring isotopes: 88, 86, 87, 84
Density: 2.54 g/cm^3 (20°C)
Melting point: 769°C   **Boiling point:** 1384°C
Latent heat of fusion: 105.1 J/g
Specific heat: 0.30 J/g/°C (25°C)
Coefficient of lineal thermal expansion: 21 × 10^{-6} cm/cm/°C (25°C)
Thermal conductivity: 0.354 w/cm/°C (25°C)
Electrical resistivity: 23 × 10^{-6} ohm-cm (20°C)
Ionization potential (1st): 5.695 eV
Electron work function ϕ: 2.59 eV
Oxidation potential: $Sr \rightarrow Sr^{2+} + 2\epsilon = 2.888$ V
Chemical valence: 2
Electrochemical equivalents: 1.635 g/amp-hr
Ionic radius: 1.12 Å (Sr^{2+})
Valence electron potential $(-\epsilon V)$: 25.7
Principal quantum number: 5
Principal electron shells: K L M N O
Electronic configuration: $1s^2\ 2s^2\ 2p^6\ 3s^2\ 3p^6\ 3d^{10}\ 4s^2\ 4p^6\ 5s^2$
Valence electrons: $5s^2$
Crystal form: Cubic, face centered
Cross section σ: 1.21 ± 0.06 barns
Vapor pressure: 2.46 × 10^2 Pa (at melting point)

Sulfur

16

32.06

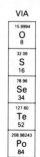

Enxôfre

Soufre

Schwefel

Azufre

cepa

גפרית

硫　硫黄

Naturally occurring isotopes: 32, 34, 33, 36
Density: 2.07 g/cm^3 (rhombic form at 25°C)
Melting point: 112.8°C   **Boiling point:** 444.674°C
Latent heat of fusion: 44.01 J/g
Specific heat: 0.706 J/g/°C (rhombic) (25°C)
Coefficient of lineal thermal expansion: 64.13 $\times$ 10^{-6} cm/cm/°C (20°C)
Thermal conductivity: 2.70 mw/cm/°C (25°C)
Electrical resistivity: 2 $\times$ 10^{17} ohm-cm (20°C)
Ionization potential (1st): 10.360 eV
Oxidation potentials: $S + 3H_2O \rightarrow H_2SO_3 + 4H^+ + 4\epsilon = -0.45$ V
$S^{2-} \rightarrow S + 2\epsilon = 0.447$ V
Chemical valence: $-2, 4, 6$
Electrochemical equivalents: 0.2990 g/amp-hr
Ionic radius: 0.37 Å (S^{4+})
Valence electron potential ($-\epsilon$V): 160
Principal quantum number: 3
Principal electron shells: K L M
Electronic configuration: $1s^2\ 2s^2\ 2p^6\ 3s^2\ 3p^4$
Valence electrons: $3s^2\ 3p^4$
Crystal form: Orthorhombic
Cross section σ: 0.51 barns
Vapor pressure: 2.65 $\times$ 10^{-20} Pa (at melting point)

Tantalum

73

180.9479

VB		
50.9415 V 23		
92.9064 Nb 41		
180.9479 Ta 73		
105		

Tantálio

Tantale

Tantal

Tántalo

тантал

טנטל

Naturally occurring isotopes: 181, 180
Density: 16.60 g/cm^3 (20°C)
Melting point: 2996°C   **Boiling point:** 5425 ± 100°C
Latent heat of fusion: 174 J/g
Specific heat: 0.140 J/g/°C (25°C)
Coefficient of lineal thermal expansion: 6.5 × 10^{-6} cm/cm/°C (20°C)
Thermal conductivity: 0.575 w/cm/°C (25°C)
Electrical resistivity: 12.45 × 10^{-6} ohm-cm (25°C)
Ionization potential (1st): 7.89 eV
Electron work function ϕ**:** 4.25 eV
Oxidation potential: $2Ta + 5H_2O \rightarrow Ta_2O_5 + 10H^+ + 10\epsilon = 0.812$ V
Chemical valence: 3, 4, *5*
Electrochemical equivalents: 1.3502 g/amp-hr
Ionic radius: 0.64 Å (Ta^{5+})
Valence electron potential ($-\epsilon$V): 110
Principal quantum number: 6
Principal electron shells: K L M N O P
Electronic configuration: $1s^2\ 2s^2\ 2p^6\ 3s^2\ 3p^6\ 3d^{10}\ 4s^2\ 4p^6\ 4d^{10}\ 4f^{14}\ 5s^2\ 5p^6$
    $5d^3\ 6s^2$
Valence electrons: 5d^3 6s^2
Crystal form: Cubic, body centered
Cross section σ**:** 22 ± 1 barns
Vapor pressure: 7.76 × 10^{-1} Pa (at melting point)

Tc Technetium

43

96.906

VIIB
54.9380 Mn 25
96.906 Tc 43
186.2 Re 75
107

Tecnécio
Technetium
Technetium
Tecnecio
технеций
טכנציום

鍀 テクネチウム

Naturally occurring isotopes: None
Density: 11.496 g/cm^3 (25°C)
Melting point: 2172°C   **Boiling point:** 4877°C
Latent heat of fusion: 235 ± 5 J/g
Specific heat: 0.24 J/g/°C (25°C)
Thermal conductivity: 0.506 w/cm/°C (25°C)
Ionization potential (1st): 7.28 eV
Oxidation potential: Tc $\rightarrow$ Tc^{2+} + 2ϵ = $-$0.4 V
Chemical valence: 0, 1, 2, 3, 4, 5, 6, 7
Electrochemical equivalents: 0.51651 g/amp-hr
Ionic radius: 0.56 Å (Tc^{7+})
Valence electron potential ($-\epsilon$V): 180
Principal quantum number: 5
Principal electron shells: K L M N O
Electronic configuration: 1s^2 2s^2 2p^6 3s^2 3p^6 3d^{10} 4s^2 4p^6 4d^6 5s^1
Valence electrons: 4d^6 5s^1
Crystal form: Hexagonal, close packed
Half life: 2.6 × 10^6 years
Vapor pressure: 2.29 × 10^{-2} Pa (at melting point)

Tellurium

52

127.60

VIA

15.9994	O — 8
32.06	S — 16
78.96	Se — 34
127.60	Te — 52
208.98243	Po — 84

Telúrio

Tellure

Tellur

Telurio

теллур

טלור

碲　テ
　　ル
　　ル

Naturally occurring isotopes: 130, 128, 126, 125, 124, 122, 123
Density: 6.24 g/cm³ (20°C)
Melting point: 449.5 ± 0.3°C   **Boiling point:** 989.8 ± 3.8°C
Latent heat of fusion: 137.2 J/g
Specific heat: 0.202 J/g/°C (25°C)
Coefficient of lineal thermal expansion: 16.75×10^{-6} cm/cm/°C (20°C)
Thermal conductivity: 0.0338 w/cm/°C (along C-axis at 25°C)
Electrical resistivity: 4.36 ohm-cm (25°C)
Ionization potential (1st): 0.009 eV
Electron work function ϕ: 4.95 eV
Oxidation potential: $Te + 2H_2O \rightarrow TeO_2 + 4H^+ + 4\epsilon = -0.529$ V
Chemical valence: $-2, 2, 4, 6$
Electrochemical equivalents: 1.1902 g/amp-hr
Ionic radius: 0.97 Å (Te^{4+})
Valence electron potential ($-\epsilon V$): 59
Principal quantum number: 5
Principal electron shells: K L M N O
Electronic configuration: $1s^2\ 2s^2\ 2p^6\ 3s^2\ 3p^6\ 3d^{10}\ 4s^2\ 4p^6\ 4d^{10}\ 5s^2\ 5p^4$
Valence electrons: $5s^2\ 5p^4$
Crystal form: Hexagonal
Cross section σ: 4.7 ± 0.1 barns
Vapor pressure: 2.31×10 Pa (at melting point)

Terbium

65

158.9254

Lanthanide Series

140.12 Ce 58	140.9077 Pr 59	144.24 Nd 60	144.913 Pm 61	150.4 Sm 62	151.96 Eu 63	157.25 Gd 64	158.9254 Tb 65	162.50 Dy 66	164.9304 Ho 67
167.26 Er 68	168.9342 Tm 69	173.04 Yb 70	174.97 Lu 71						

Térbio
Terbium
Terbium
Terbio
тербий
טרביום

鉽 テルビウム

Naturally occurring isotope: 159
Density: 8.229 g/cm^3 (25°C)
Melting point: 1356°C   **Boiling point:** 3123°C
Latent heat of fusion: 102.7 J/g
Specific heat: 0.182 J/g/°C (25°C)
Coefficient of lineal thermal expansion: 11.8 × 10^{-6} cm/cm/°C (25°C)
Thermal conductivity: 0.111 w/cm/°C (25°C)
Electrical resistivity: 116 × 10^{-6} ohm-cm (25°C)
Ionization potential (1st): 5.85 eV
Electron work function φ: 3.0 eV
Oxidation potential: Tb → Tb^{3+} + 3ε = 2.391 V
Chemical valence: *3,* 4
Electrochemical equivalents: 1.9765 g/amp-hr
Ionic radius: 0.923 Å (Tb^{3+})
Valence electron potential (−εV): 46.8
Principal quantum number: 6
Principal electron shells: K L M N O P
Electronic configuration: 1s^2 2s^2 2p^6 3s^2 3p^6 3d^{10} 4s^2 4p^6 4d^{10} 4f^9 5s^2
    5p^6 6s^2
Valence electrons: 4f^9 6s^2
Crystal form: Hexagonal, close packed
Cross section σ: 30 ± 10 barns

Tl **Thallium**

81

204.37

IIIA

| 10.81 |
| B |
| 5 |

| 26.98154 |
| Al |
| 13 |

| 69.72 |
| Ga |
| 31 |

| 114.82 |
| In |
| 49 |

| 204.37 |
| Tl |
| 81 |

Tálio
Thallium
Thallium
Talio
таллий
תליום
鉈 タリウム

Naturally occurring isotopes: 205, 203
Density: 11.85 g/cm^3 (20°C)
Melting point: 303.5°C   **Boiling point:** 1457 ± 10°C
Latent heat of fusion: 20.90 J/g
Specific heat: 0.129 J/g/°C (25°C)
Coefficient of lineal thermal expansion: 28 × 10^{-6} cm/cm/°C (20°C)
Thermal conductivity: 0.461 w/cm/°C (25°C)
Electrical resistivity: 18.0 × 10^{-6} ohm-cm (0°C)
Ionization potential (1st): 6.108 eV
Electron work function φ: 3.84 eV
Oxidation potentials: Tl → Tl$^+$ + ε = 0.3363 V
$\qquad\qquad$ Tl$^+$ → Tl^{3+} + 2ε = −1.25 V
Chemical valence: *1,* 3
Electrochemical equivalents: 7.6250 g/amp-hr
Ionic radius: 1.50 Å (Tl$^+$)
Valence electron potential (−εV): 9.60
Principal quantum number: 6
Principal electron shells: K L M N O P
Electronic configuration: 1s^2 2s^2 2p^6 3s^2 3p^6 3d^{10} 4s^2 4p^6 4d^{10} 4f^{14} 5s^2 5p^6
$\qquad$ 5d^{10} 6s^2 6p^1
Valence electrons: 6s^2 6p^1
Crystal form: Hexagonal, close packed
Cross section σ: 3.4 ± 0.5 barns
Vapor pressure: 5.33 × 10^{-6} Pa (at melting point)

Thorium

90

232.03807

Actinide Series

Tório
Thorium
Thorium
Torio
торий
תוריום

232.03807	231.0359	238.029	237.0482	244.06423	243.0614	247.07038	247.07032	251.07961	254.08805
Th	Pa	U	Np	Pu	Am	Cm	Bk	Cf	Es
90	91	92	93	94	95	96	97	98	99
257.09515	258	259	260						
Fm	Md	No	Lr						
100	101	102	103						

釷 トリウム

Naturally occurring isotope: 232
Density: 11.724 g/cm^3 (25°C)
Melting point: 1750°C   **Boiling point:** 4787°C
Latent heat of fusion: 82.93 J/g
Specific heat: 0.118 J/g/°C (25°C)
Coefficient of lineal thermal expansion: 12.5 × 10^{-6} cm/cm/°C (20°C)
Thermal conductivity: 0.540 w/cm/°C (25°C)
Electrical resistivity: 13.1 × 10^{-6} ohm-cm (25°C)
Ionization potential (1st): 6.08 eV
Electron work function ϕ: 3.41 eV
Oxidation potential: Th → Th^{4+} + 4ϵ = 1.899 V
Chemical valence: 3, *4*
Electrochemical equivalents: 2.1643 g/amp-hr
Ionic radius: 0.972 Å (Th^{4+})
Valence electron potential (−εV): 59.3
Principal quantum number: 7
Principal electron shells: K L M N O P Q
Electronic configuration: 1s^2 2s^2 2p^6 3s^2 3p^6 3d^{10} 4s^2 4p^6 4d^{10} 4f^{14} 5s^2 5p^6
    5d^{10} 6s^2 6p^6 6d^2 7s^2
Valence electrons: 6d^2 7s^2
Crystal form: Cubic, face centered
Half life: 1.40 × 10^{10} years
Cross section σ: 74±0.1 barns

Thulium

69

168.9342

Lanthanide Series

140.12 Ce 58	140.9077 Pr 59	144.24 Nd 60	144.913 Pm 61	150.4 Sm 62	151.96 Eu 63	157.25 Gd 64	158.9254 Tb 65	162.50 Dy 66	164.9304 Ho 67
167.26 Er 68	168.9342 Tm 69	173.04 Yb 70	174.97 Lu 71						

Túlio
Thulium
Thulium
Tulio
тулий
תוליום

銩 ツリウム

Naturally occurring isotope: 169
Density: 9.321 g/cm^3 (25°C)
Melting point: 1545 ± 15°C   **Boiling point:** 1727°C
Latent heat of fusion: 109.0 J/g
Specific heat: 0.160 J/g/°C (25°C)
Coefficient of lineal thermal expansion: 11.6 × 10^{-6} cm/cm/°C (400°C)
Thermal conductivity: 0.169 w/cm/°C (25°C)
Electrical resistivity: 79 × 10^{-6} ohm-cm (25°C)
Ionization potential (1st): 6.1844 eV
Oxidation potential: Tm → Tm^{3+} + 3ϵ = 2.278 V
Chemical valence: 2, *3*
Electrochemical equivalents: 2.1010 g/amp-hr
Ionic radius: 0.869 Å (Tm^{3+})
Valence electron potential ($-\epsilon$V): 49.7
Principal quantum number: 6
Principal electron shells: K L M N O P
Electronic configuration: 1s^2 2s^2 2p^6 3s^2 3p^6 3d^{10} 4s^2 4p^6 4d^{10} 4f^{13} 5s^2 5p^6 6s^2
Valence electrons: 4f^{13} 6s^2
Crystal form: Hexagonal, close packed
Cross section σ: 115 ± 15 barns
Vapor pressure: 4.90 × 10^{-3} Pa (at melting point)

Tin

50

118.69

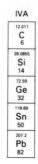

IVA

12.011 C 6
28.0855 Si 14
72.59 Ge 32
118.69 Sn 50
207.2 Pb 82

Estanho

Etain

Zinn

Estaño

олово

בדיל

錫 す ず

Naturally occurring isotopes: 120, 118, 116, 119, 117, 124, 122, 112, 114, 115

Density: 7.298 g/cm^3 (25°C)

Melting point: 231.9681°C   **Boiling point:** 2270°C

Latent heat of fusion: 60.67 J/g

Specific heat: 0.227 J/g/°C (25°C)

Coefficient of lineal thermal expansion: 23 × 10^{-6} cm/cm/°C (20°C)

Thermal conductivity: 0.668 w/cm/°C (25°C)

Electrical resistivity: 11.5 × 10^{-6} ohm-cm (20°C)

Ionization potential (1st): 7.334 eV

Electron work function ϕ: 4.42 eV

Oxidation potentials: Sn → Sn^{2+} + 2ϵ = 0.136 V

Sn^{2+} → Sn^{4+} + 2ϵ = −0.15 V

Chemical valence: −4, −1, 2, *4*

Electrochemical equivalents: 1.1071 g/amp-hr

Ionic radius: 0.690 Å (Sn^{4+})

Valence electron potential (−ϵV): 83.5

Principal quantum number: 5

Principal electron shells: K L M N O

Electronic configuration: 1s^2 2s^2 2p^6 3s^2 3p^6 3d^{10} 4s^2 4p^6 4d^{10} 5s^2 5p^2

Valence electrons: 5s^2 5p^2

Crystal form: Tetragonal

Cross section σ: 0.63 ± 0.1 barns

Vapor pressure: 5.78 × 10^{-21} Pa (at melting point)

Ti Titanium

22

47.90

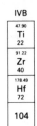

Titânio
Titane
Titan
Titanio
титан

סיטניום

鈦 チタン

Naturally occurring isotopes: 48, 46, 47, 49, 50
Density: 4.507 g/cm^3 (20°C)
Melting point: 1660 ± 10°C   **Boiling point:** 3287°C
Latent heat of fusion: 323.4 J/g
Specific heat: 0.522 J/g/°C (25°C)
Coefficient of lineal thermal expansion: 8.41 × 10^{-6} cm/cm/°C (20°C)
Thermal conductivity: 0.219 w/cm/°C (25°C)
Electrical resistivity: 42 × 10^{-6} ohm-cm (20°C)
Ionization potential (1st): 6.82 eV
Electron work function ϕ: 4.33 eV
Oxidation potential: Ti → Ti^{2+} + 2ϵ = 1.628 V
Chemical valence: 1, 2, 3, *4*
Electrochemical equivalents: 0.4468 g/amp-hr
Ionic radius: 0.605 Å (Ti^{4+})
Valence electron potential ($-\epsilon$V): 95.2
Principal quantum number: 4
Principal electron shells: K L M N
Electronic configuration: 1s^2 2s^2 2p^6 3s^2 3p^6 3d^2 4s^2
Valence electrons: 3d^2 4s^2
Crystal form: Hexagonal, close packed
Cross section σ: 6.1 ± 0.2 barns
Vapor pressure: 4.90 × 10^{-1} Pa (at melting point)

IVB

| 47.90 |
| Ti |
| 22 |

| 91.22 |
| Zr |
| 40 |

| 178.49 |
| Hf |
| 72 |

| 104 |

W Tungsten

74

183.85

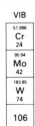

	VIB
	51.996
	Cr
	24
	95.94
	Mo
	42
	183.85
	W
	74
	106

Tungstênio

Tungstène

Wolframz

Tungsteno

вольфрам

וולפרם

鎢 タングステン

Naturally occurring isotopes: 184, 186, 182, 183, 180
Density: 19.35 g/cm³ (20°C)
Melting point: 3410 ± 20°C   **Boiling point:** 5660°C
Latent heat of fusion: 191.7 J/g
Specific heat: 0.125 J/g/°C (25°C)
Coefficient of lineal thermal expansion: 4.6×10^{-6} cm/cm/°C (20°C)
Thermal conductivity: 1.73 w/cm/°C (25°C)
Electrical resistivity: 5.65×10^{-6} ohm-cm (27°C)
Ionization potential (1st): 7.98 eV
Electron work function ϕ: 4.55 eV
Oxidation potential: $W + 3H_2O \rightarrow WO_3 + 6H^+ + 6\epsilon = 0.09$ V
Chemical valence: 2, 3, 4, 5, *6*
Electrochemical equivalents: 1.1432 g/amp-hr
Ionic radius: 0.62 Å (W^{6+})
Valence electron potential ($-\epsilon V$): 140
Principal quantum number: 6
Principal electron shells: K L M N O P
Electronic configuration: $1s^2\ 2s^2\ 2p^6\ 3s^2\ 3p^6\ 3d^{10}\ 4s^2\ 4p^6\ 4d^{10}\ 4f^{14}\ 5s^2\ 5p^6$ $5d^4\ 6s^2$
Valence electrons: $5d^4\ 6s^2$
Crystal form: Alpha—cubic, body centered; beta—cubic, face centered
Cross section σ: 18.5 ± 0.5 barns
Vapor pressure: 4.27 Pa (at melting point)

Uranium

92

238.029

Actinide Series

232.03807	231.0359	238.029	237.0482	244.06423	243.0614	247.07038	247.07032	251.07961	254.08805
Th	Pa	U	Np	Pu	Am	Cm	Bk	Cf	Es
90	91	92	93	94	95	96	97	98	99
257.09515	258	259	260						
Fm	Md	No	Lr						
100	101	102	103						

Urânio
Uranium
Uran
Uranio
уран

אורניום

鈾 ウラニウム

Naturally occurring isotopes: 238, 235, 234
Density: 19.04 g/cm^3 (25°C)
Melting point: 1132.3 ± 0.8°C   **Boiling point:** 3818°C
Latent heat of fusion: 65.08 J/g
Specific heat: 0.1162 J/g/°C (25°C)
Coefficient of lineal thermal expansion: 13.4 × 10^{-6} cm/cm/°C (25°C)
Thermal conductivity: 0.275 w/cm/°C (25°C)
Electrical resistivity: 27 × 10^{-6} ohm-cm (25°C)
Ionization potential (1st): 6.05 eV
Electron work function ϕ: 3.63 eV
Oxidation potential: $U \rightarrow U^{3+} + 3\epsilon = 1.789$ V
Chemical valence: 3, 4, 5, *6*
Electrochemical equivalents: 1.4801 g/amp-hr
Ionic radius: 0.52 Å (U^{6+})
Valence electron potential (−ϵV): 170
Principal quantum number: 7
Principal electron shells: K L M N O P Q
Electronic configuration: $1s^2$ $2s^2$ $2p^6$ $3s^2$ $3p^6$ $3d^{10}$ $4s^2$ $4p^6$ $4d^{10}$ $4f^{14}$ $5s^2$ $5p^6$
    $5d^{10}$ $5f^3$ $6s^2$ $6p^6$ $6d^1$ $7s^2$
Valence electrons: $5f^3$ $6d^1$ $7s^2$
Crystal form: Orthorhombic
Half life: 4.51 × 10^9 years
Cross section σ: 7.595 ± 0.070 barns
Vapor pressure: 1.19 × 10^{-6} Pa (at melting point)

Vanadium

23

50.9415

VB	
50.9415	
V	
23	
92.9064	
Nb	
41	
180.9479	
Ta	
73	
105	

Vanádio
Vanadium
Vanadium
Vanadio
ванадий

ונדיום

釩 バナジウム

Naturally occurring isotopes: 51, 50
Density: 6.11 g/cm^3 (18.7°C)
Melting point: 1890 ± 10°C   **Boiling point:** 3380°C
Latent heat of fusion: 345.2 J/g
Specific heat: 0.489 J/g/°C (25°C)
Coefficient of lineal thermal expansion: 6.15 × 10^{-6} cm/cm/°C (25°C)
Thermal conductivity: 0.307 w/cm/°C (25°C)
Electrical resistivity: 24.8 × 10^{-6} ohm-cm (20°C)
Ionization potential (1st): 6.74 eV
Electron work function φ: 4.3 eV
Oxidation potential: $V \rightarrow V^{2+} + 2\epsilon = 1.186$ V
Chemical valence: 2, 3, 4, *5*
Electrochemical equivalents: 0.38013 g/amp-hr
Ionic radius: 0.59 Å (V^{5+})
Valence electron potential (−εV): 120
Principal quantum number: 4
Principal electron shells: K L M N
Electronic configuration: $1s^2\ 2s^2\ 2p^6\ 3s^2\ 3p^6\ 3d^3\ 4s^2$
Valence electrons: $3d^3\ 4s^2$
Crystal form: Cubic, body centered
Cross section σ: 5.06 ± 0.06 barns
Vapor pressure: 3.06 Pa (at melting point)

Xenon

54

131.30

	O
	4.00260 **He** 2
	20.179 **Ne** 10
	39.948 **Ar** 18
	83.80 **Kr** 36
	131.30 **Xe** 54
	222.01761 **Rn** 86

Xenônio

Xènon

Xenon

Xenón

ксенон

כסנית

氙 キセノン

Naturally occurring isotopes: 132, 129, 131, 134, 136, 130, 128, 124, 126

Density: 5.895×10^{-3} g/cm^3 (20°C)

Melting point: -111.9°C   Boiling point: -107.1 ± 3°C

Latent heat of fusion: 17.5 J/g

Specific heat: 0.15831 J/g/°C (25°C)

Thermal conductivity: 0.514 mw/cm/°C (0°C at 1 atm)

Ionization potential (1st): 12.130 eV

Chemical valence: 0

Principal quantum number: 5

Principal electron shells: K L M N O

Electronic configuration: $1s^2\ 2s^2\ 2p^6\ 3s^2\ 3p^6\ 3d^{10}\ 4s^2\ 4p^6\ 4d^{10}\ 5s^2\ 5p^6$

Valence electrons: $(5s^2\ 5p^6)$

Crystal form: Cubic, face centered

Cross section σ: 24.5 ± 1.0 barns

Yb Ytterbium

70

173.04

Lanthanide Series

140.12 Ce 58	140.9077 Pr 59	144.24 Nd 60	144.913 Pm 61	150.4 Sm 62	151.96 Eu 63	157.25 Gd 64	158.9254 Tb 65	162.50 Dy 66	164.9304 Ho 67
167.26 Er 68	168.9342 Tm 69	173.04 Yb 70	174.97 Lu 71						

Itérbio
Ytterbium
Ytterbium
Iterbio
иттербий
איטרביום

 イッテルビウム

Naturally occurring isotopes: 174, 172, 173, 171, 176, 170, 168
Density: 6.965 g/cm^3 (25°C)
Melting point: 819°C   **Boiling point:** 1194°C
Latent heat of fusion: 53.23 J/g
Specific heat: 0.155 J/g/°C (25°C)
Coefficient of lineal thermal expansion: 29.9 $\times$ 10^6 cm/cm/°C (25°C)
Thermal conductivity: 0.349 w/cm/°C (25°C)
Electrical resistivity: 28 $\times$ 10^{-6} ohm-cm (25°C)
Ionization potential (1st): 6.2539 eV
Oxidation potential: Yb $\rightarrow$ Yb^{3+} + 3ϵ = 2.267 V
Chemical valence: 2, *3*
Electrochemical equivalents: 2.1520 g/amp-hr
Ionic radius: 0.858 Å (Yb^{3+})
Valence electron potential ($-\epsilon$V): 50.3
Principal quantum number: 6
Principal electron shells: K L M N O P
Electronic configuration: 1s^2 2s^2 2p^6 3s^2 3p^6 3d^{10} 4s^2 4p^6 4d^{10} 4f^{14} 5s^2
5p^6 6s^2
Valence electrons: 4f^{14} 6s^2
Crystal form: Cubic, face centered
Cross section σ: 37 ± 3 barns
Vapor pressure: 3.95 $\times$ 10^2 Pa (at melting point)

Y Yttrium

39

88.9059

IIIB
44.95592 Sc 21
88.9059 Y 39
138.9055 La 57
227.02777 Ac 89

Itrio
Yttrium
Yttrium
Itrio
иттрий
איטריום

 イットリウム

Naturally occurring isotope: 89
Density: 4.469 g/cm³ (25°C)
Melting point: 1522°C   **Boiling point:** 3338°C
Latent heat of fusion: 193.1 J/g
Specific heat: 0.298 J/g/°C (25°C)
Coefficient of lineal thermal expansion: 10.8×10^{-6} cm/cm/°C (400°C)
Thermal conductivity: 0.172 w/cm/°C (25°C)
Electrical resistivity: 57×10^{-6} ohm-cm (25°C)
Ionization potential (1st): 6.38 eV
Electron work function ϕ: 3.1 eV
Oxidation potential: $Y \rightarrow Y^{3+} + 3\epsilon = 2.372$ V
Chemical valence: 3
Electrochemical equivalents: 1.1057 g/amp-hr
Ionic radius: 0.900 Å (Y^{3+})
Valence electron potential $(-\epsilon V)$: 48.0
Principal quantum number: 5
Principal electron shells: K L M N O
Electronic configuration: $1s^2 \, 2s^2 \, 2p^6 \, 3s^2 \, 3p^6 \, 3d^{10} \, 4s^2 \, 4p^6 \, 4d^1 \, 5s^2$
Valence electrons: $4d^1 \, 5s^2$
Crystal form: Hexagonal, close packed
Cross section σ: 1.3 ± 0.1 barns
Vapor pressure: 5.31 Pa (at melting point)

Zn **Zinc**

30	65.38 **Zn** 30
65.38	112.41 **Cd** 48
	200.59 **Hg** 80

Zinco

Zinc

Zink

Zinc

цинк

אבץ

鋅 亜鉛

Naturally occurring isotopes: 64, 66, 68, 67, 70
Density: 7.133 g/cm^3 (25°C)
Melting point: 419.58°C   **Boiling point:** 907°C
Latent heat of fusion: 113.0 J/g
Specific heat: 0.388 J/g/°C (25°C)
Coefficient of lineal thermal expansion: 39.7 $\times$ 10^{-6} cm/cm/°C (20°C)
Thermal conductivity: 1.16 w/cm/°C (25°C)
Electrical resistivity: 5.916 $\times$ 10^{-6} ohm-cm (20°C)
Ionization potential (1st): 9.394 eV
Electron work function ϕ: 4.33 eV
Oxidation potential: Zn $\rightarrow$ Zn^{2+} + 2ϵ = 0.7628 V
Chemical valence: 2
Electrochemical equivalents: 1.220 g/amp-hr
Ionic radius: 0.740 Å (Zn^{2+})
Valence electron potential ($-\epsilon$V): 38.9
Principal quantum number: 4
Principal electron shells: K L M N
Electronic configuration: 1s^2 2s^2 2p^6 3s^2 3p^6 3d^{10} 4s^2
Valence electrons: 4s^2
Crystal form: Hexagonal, close packed
Cross section σ: 1.10 ± 0.04 barns
Vapor pressure: 19.2 Pa (at melting point)

Zr **Zirconium**

40

91.22

Zircônio
Zirconium
Zirkonium
Zirconio
цирконий
צירקוניום

鋯 ジルコニウム

Naturally occurring isotopes: 90, 94, 92, 91, 96
Density: 6.506 g/cm^3 (20°C)
Melting point: 1852 ± 2°C   **Boiling point:** 4377°C
Latent heat of fusion: 251.2 J/g
Specific heat: 0.278 J/g/°C (25°C)
Coefficient of lineal thermal expansion: 5.85 × 10^{-6} cm/cm/°C (20°C)
Thermal conductivity: 0.227 w/cm/°C (27°C)
Electrical resistivity: 40 × 10^{-6} ohm-cm (20°C)
Ionization potential (1st): 6.84 eV
Electron work function ϕ: 4.05 eV
Oxidation potential: Zr → Zr^{4+} + 4ϵ = 1.529 V
Chemical valence: 1, 2, 3, *4*
Electrochemical equivalents: 0.8509 g/amp-hr
Ionic radius: 0.72 Å (Zr^{4+})
Valence electron potential ($-\epsilon$V): 80
Principal quantum number: 5
Principal electron shells: K L M N O
Electronic configuration: 1s^2 2s^2 2p^6 3s^2 3p^6 3d^{10} 4s^2 4p^6 4d^2 5s^2
Valence electrons: 4d^2 5s^2
Crystal form: Hexagonal, close packed
Cross section σ: 0.182 ± 0.005 barns
Vapor pressure: 1.68 × 10^{-3} Pa (at melting point)

Kurchatovium
Rutherfordium

104

261

IVB
47.90 **Ti** 22
91.22 **Zr** 40
178.49 **Hf** 72
104

Naturally occurring isotopes: None

Chemical valence: 4

Principal quantum number: 7

Principal electron shells: K L M N O P Q

Electronic configuration: $1s^2 2s^2 2p^6 3s^2 3p^6 3d^{10} 4s^2 4p^6 4d^{10} 4f^{14} 5s^2 5p^6$ $5d^{10} 5f^{14} 6s^2 6p^6 6d^2 7s^2$

Valence electrons: $6d^2 7s^2$

Half life: ~65 seconds

Nielsbohrium
Hahnium

105

(262)

VB
50.9415   **V**   23
92.9064   **Nb**   41
180.9479   **Ta**   73
105

Naturally occurring isotopes: None
Chemical valence: (5)
Principal quantum number: 7
Principal electron shells: K L M N O P Q
Half life: ~40 seconds

Element
106

106

(263)

VIB
51.996
Cr
24
95.94
Mo
42
183.85
W
74
106

Naturally occurring isotopes: None
Chemical valence: (6)
Principal quantum number: 7
Principal electron shells: K L M N O P Q
Half life: ~1 second

Element

107
(262)

VIIB
54.9380 Mn 25
96.906 Tc 43
186.2 Re 75
107

Naturally occurring isotopes: None
Chemical valence: (7)
Principal quantum number: 7
Principal electron shells: K L M N O P Q

Element

109

(266)

VIII		
55.847 **Fe** 26	58.9332 **Co** 27	58.70 **Ni** 28
101.07 **Ru** 44	102.9055 **Rh** 45	106.4 **Pd** 46
190.2 **Os** 76	192.22 **Ir** 77	195.09 **Pt** 78
	109	

Naturally occurring isotopes: None
Chemical valence: (3)
Principal quantum number: 7
Principal electron shells: K L M N O P Q

Bibliography

Of the numerous references employed in preparation of the third edition of the "Handbook of the Elements," the following are among the most prominent:

A. E. Bailey et al., Eds., "Tables of Physical and Chemical Constants," 14th ed., Longman Group Limited, London (1973).

I. Barin and O. Knacke, "Thermochemical Properties of Inorganic Substances," Springer-Verlag, Berlin (1973).

G. Charlot et al., "Selected Constants: Oxidation–Reduction Potentials of Inorganic Substances in Aqueous Solution," Butterworths, London (1971).

J. A. Dean, Ed., "Lange's Handbook of Chemistry," 12th ed., McGraw-Hill Book Company, New York (1978).

S. Fraga and J. Karwowski, "Handbook of Atomic Data," Elsevier Scientific Publishing Company, Amsterdam (1976).

"Gmelin Handbuch der Anorganischen Chemie," Springer-Verlag, Berlin (up to and including 1980).

M. Grayson, Executive Ed., "Kirk-Othmer Encyclopedia of Chemical Technology," 3rd ed., Volumes 1–20 incl., John Wiley & Sons, New York (1978–1982 incl.)

G. B. Naumov et al., Eds., "Handbook of Thermodynamic Data," U.S.S.R. Academy of Sciences, Leningrad (1971).

L. Pauling, "The Nature of the Chemical Bond," 3rd ed., Cornell University Press, Ithaca, New York (1960).

G. V. Samsonov, "Handbook of the Physicochemical Properties of the Elements," Plenum, New York (1968).

R. C. Weast, Editor-in-Chief, "CRC Handbook of Chemistry and Physics, 1983–1984," 64th ed., CRC Press, Inc., Boca Raton, Fla. (1983).

E. Ya. Zandberg and N. L. Ionov, "Surface Ionization," Science Publishing House, Moscow (1969).

In addition to the reference texts, several primary journals and U.S. government publications were employed. The most commonly utilized were:

Acta Chemica Scandinavica
Acta Crystallographica
Analytical Chemistry
Bulletin of the American Physical Society
Canadian Journal of Chemistry
Canadian Journal of Physics
Chemical and Engineering News
Chemical Physics Letters
Chemical Reviews
Chemische Berichte
Electrochimica Acta

Helvetica Chimica Acta
Inorganic Chemistry
Journal of American Chemical Society
Journal of Applied Physics
Journal of Chemical and Engineering Data
Journal of Chemical Education
Journal of Chemical Physics
Journal of Inorganic and Nuclear Chemistry
Journal of Less-Common Metals
Journal of Physical and Chemical Reference Data
Journal of Physical Chemistry
Journal of Physics and Chemistry of Solids
Journal of Solid State Chemistry
Journal of the Chemical Society
Journal of the Electrochemical Society
Materials Research Bulletin
Nature
Physical Review
Proceedings of the Physical Society
Proceedings of the Royal Society
Progress in Inorganic Chemistry
Pure and Applied Chemistry
Science
Talanta
Transactions of the Faraday Society